Springer Theses

Recognizing Outstanding Ph.D. Research

Aims and Scope

The series "Springer Theses" brings together a selection of the very best Ph.D. theses from around the world and across the physical sciences. Nominated and endorsed by two recognized specialists, each published volume has been selected for its scientific excellence and the high impact of its contents for the pertinent field of research. For greater accessibility to non-specialists, the published versions include an extended introduction, as well as a foreword by the student's supervisor explaining the special relevance of the work for the field. As a whole, the series will provide a valuable resource both for newcomers to the research fields described, and for other scientists seeking detailed background information on special questions. Finally, it provides an accredited documentation of the valuable contributions made by today's younger generation of scientists.

Theses are accepted into the series by invited nomination only and must fulfill all of the following criteria

- They must be written in good English.
- The topic should fall within the confines of Chemistry, Physics, Earth Sciences, Engineering and related interdisciplinary fields such as Materials, Nanoscience, Chemical Engineering, Complex Systems and Biophysics.
- The work reported in the thesis must represent a significant scientific advance.
- If the thesis includes previously published material, permission to reproduce this must be gained from the respective copyright holder.
- They must have been examined and passed during the 12 months prior to nomination.
- Each thesis should include a foreword by the supervisor outlining the significance of its content.
- The theses should have a clearly defined structure including an introduction accessible to scientists not expert in that particular field.

More information about this series at http://www.springer.com/series/8790

Dominic Rüfenacht

Novel Motion Anchoring Strategies for Wavelet-Based Highly Scalable Video Compression

Doctoral Thesis accepted by
the UNSW Sydney, Australia

Springer

Author
Dr. Dominic Rüfenacht
Electrical Engineering and
  Telecommunications
UNSW Sydney
Sydney, NSW
Australia

Supervisor
Prof. David Taubman
Electrical Engineering and
  Telecommunications
UNSW Sydney
Sydney, NSW
Australia

ISSN 2190-5053   ISSN 2190-5061 (electronic)
Springer Theses
ISBN 978-981-13-4097-0   ISBN 978-981-10-8225-2 (eBook)
https://doi.org/10.1007/978-981-10-8225-2

Printed on acid-free paper

This Springer imprint is published by Springer Nature
The registered company is Springer Nature Singapore Pte Ltd.
The registered company address is: 152 Beach Road, #21-01/04 Gateway East, Singapore 189721, Singapore

Supervisor's Foreword

Throughout the history of digital video coding, motion information has played a critical role, allowing the effective exploitation of temporal redundancy. Not surprisingly, advances in coding technology have been substantially driven by improvements in motion modeling, motion coding and inter-frame motion compensation techniques. Virtually without exception, the motion used for video coding has been modeled, estimated, and coded as a vector field that is anchored at the (target) frames that the motion will be used to predict, based on other (reference) frames. This target-anchored approach is compelling, because it ensures that every sample to be predicted has a motion-induced association with exactly one location in the corresponding reference frame, ignoring the constraints imposed by frame boundaries, which facilitates motion compensation and estimation tasks.

In this thesis by Dominic Rüfenacht, a remarkably strong case is presented for revisiting the target-based motion anchoring convention. A key theme in the argument is temporal consistency. Target-anchored motion provides a convenient heuristic for motion-compensated prediction of the target frame, but cannot easily be repurposed for the prediction or synthesis of other frames. This is a problem for temporal frame interpolation, and also for scalable video coding. With multiple reference frames, target-anchored motion provides multiple ways to infer new motion relationships, and these are not generally consistent. This thesis proposes to solve these problems by anchoring motion information at the reference frames instead. It is shown that this can lead to improved coding efficiency, as well as superior visual quality when motion fields are degraded.

Central to the success of a reference-anchored approach is high-quality temporal motion inference, allowing motion fields to be inverted and composed while preserving motion discontinuities. This, in turn, relies upon the estimation and coding of piecewise-smooth motion fields which preserve motion boundaries, the primary sources of innovation in a video sequence. The publication of this thesis is particularly timely, because motion estimation and coding techniques are now reaching the point where they are able to meet these prerequisites.

Since temporal frame interpolation (TFI) is both an important element in most modern display devices and a close proxy for motion-compensated prediction in video coding, the thesis also explores the potential of the reference-anchored approach in a TFI setting. Remarkably, the proposed reference-anchored motion inference strategy is found to outperform state-of-the-art TFI schemes.

The novel motion anchoring strategies proposed in this thesis are well adapted to supporting "features" beyond compression, including high scalability, accessibility, and "intrinsic" frame upsampling. These are important to the transition from a frame-centric broadcast paradigm to a future in which video content is interactively browsed and repurposed within heterogeneous networks and systems.

Sydney, Australia   Prof. David Taubman
August 2017

Abstract

This thesis investigates new motion anchoring strategies that are targeted at wavelet-based highly scalable video compression (WSVC), which represents a fundamental change to the way motion information is employed in a video compression system. We depart from two practices that are deeply ingrained in existing video compression systems. Instead of the commonly used block motion, which has poor scalability attributes, we employ piecewise-smooth motion together with a highly scalable motion boundary description. The combination of this more "physical" motion description together with motion-discontinuity information allows us to change the conventional strategy of anchoring motion at target frames to anchoring motion at reference frames, which improves motion inference across time.

In the proposed *reference-based* motion anchoring strategies, motion fields are mapped from reference to target frames, where they serve as prediction references; during this mapping process, disoccluded regions are readily discovered. Observing that motion discontinuities displace with foreground objects, we propose motion-discontinuity-driven motion mapping operations that handle traditionally challenging regions around moving objects. The reference-based motion anchoring exposes an intricate connection between temporal frame interpolation (TFI) and video compression. When employed in a compression system, all anchoring strategies explored in this thesis perform TFI once all residual information is quantized to zero at a given temporal level. The interpolation performance is evaluated on both natural and synthetic sequences, where we show favorable comparisons with state-of-the-art TFI schemes.

We explore three reference-based motion anchoring strategies. In the first one, the motion anchoring is "flipped" with respect to a hierarchical B-frame structure. We develop an analytical model to determine the weights of the different spatiotemporal subbands, and assess the suitability and benefits of this reference-based WSVC for (highly scalable) video compression. Reduced motion coding cost and improved frame prediction, especially around moving objects, result in improved rate–distortion performance compared to a target-based WSVC. As the thesis evolves, the motion anchoring is progressively simplified to one where all motion is anchored at one base frame; this central motion organization facilitates the incorporation of higher order motion models, which improve the *prediction* performance in regions following motion with nonconstant velocity.

This novel motion anchoring paradigm is well adapted to seamlessly supporting "features" beyond compressibility, including high scalability, accessibility, and "intrinsic" frame upsampling. These features are becoming ever more relevant as the way video is consumed continues shifting from the traditional broadcast scenario with predefined network and decoder constraints to *interactive browsing* of video content over heterogeneous networks.

Parts of this thesis have been published in the following articles:

- [1] D. Rüfenacht, R. Mathew, and D. Taubman. "Hierarchical Anchoring of Motion Fields for Fully Scalable Video Coding". *IEEE International Conference on Image Processing (ICIP)*, 2014.
- [2] D. Rüfenacht, R. Mathew, and D. Taubman. "Bidirectional, Occlusion-Aware Temporal Frame Interpolation in a Highly Scalable Video Setting". *Picture Coding Symposium (PCS)*, 2015.
- [3] D. Rüfenacht, R. Mathew, and D. Taubman. "Occlusion-Aware Temporal Frame Interpolation in a Highly Scalable Video Coding Setting". *APSIPA Transactions on Signal and Information Processing (ATSIP)*, vol. 5, 2016.
- [4] D. Rüfenacht, R. Mathew, and D. Taubman. "Bidirectional Hierarchical Anchoring of Motion Fields for Scalable Video Coding". *IEEE International Workshop on Multimedia Signal Processing (MMSP)*, 2014. **Top 10% Award**.
- [5] D. Rüfenacht, R. Mathew, and D. Taubman. "Motion Blur Modelling for Hierarchically Anchored Motion with Discontinuities". *IEEE International Workshop on Multimedia Signal Processing (MMSP)*, 2015.
- [6] D. Rüfenacht, R. Mathew, and D. Taubman. "A Novel Motion Field Anchoring Paradigm for Highly Scalable Wavelet-based Video Coding". *IEEE Transactions on Image Processing*, vol. 25, no. 1, pp. 39–52, 2016.
- [7] D. Rüfenacht, R. Mathew, and D. Taubman. "Temporal Frame Interpolation with Motion-divergence-guided Occlusion Handling". *IEEE Transactions on Circuits and Systems for Video Technology*, 2018 (in press).
- [8] D. Rüfenacht and D. Taubman. "Temporally Consistent High Frame-Rate Upsampling with Motion Sparsifcation". *IEEE International Workshop on Multimedia Signal Processing (MMSP)*, 2016.
- [9] D. Rüfenacht, R. Mathew, and D. Taubman. "Higher-Order Motion Models for Temporal Frame Interpolation with Applications to Video Coding". *Picture Coding Symposium (PCS)*, 2016.

Acknowledgements

It's always further than it looks. It's always taller than it looks. And it's always harder than it looks.

Looking back at my time as a Ph.D. candidate, the *three rules of mountaineering* resonate with me. I would like to thank a number of people who helped me reach the top of the mountain – the view is beautiful.

First and foremost, my most sincere thanks and appreciation go to my supervisor, Prof. David Taubman, who made sure that my backpack was always full of ideas. Thank you for your excellent guidance, advice, and support throughout my candidature.

Special thanks go to Dr. Reji Mathew, for his technical and moral support throughout the thesis. I wish to extend my thanks to the other members of the Interactive Visual Media Processing (IVMP) Lab – Dr. Aous Thabit Naman, Dr. Rui Xu, Sean Young, and Maryam Haghighat – for their friendship and insightful discussions over coffee and lunch breaks. A number of people outside the lab made my Ph.D. a more pleasant one. I particularly enjoyed the culinary excursions with Nicholas Cummins, Stefanie Brown, Shanglin Ye, Sanat Biswas, Vidhya Sethu, and Tom Millet. I owe a great deal of thanks to Prof. Sabine Süsstrunk from EPFL – for sparking my interest in research, patiently helping me write my first academic papers, and for establishing the initial contact with my thesis supervisor.

I acknowledge UNSW Australia for the Tuition Fee Scholarship (TFS) and a Faculty Research Stipend, which allowed me to undertake the University's doctoral program. I would also like to thank my supervisor for his generous support with publications and conference-related travel.

I am grateful to my parents for their unconditional love and understanding, and for making sure I had the best possible education. I would also like to thank my extended family for all their support, in particular, for all the Swiss treats reinforcement packages that made climbing the mountain this much easier. Finally, the most heartfelt thanks go to my lifelong travel companion Simone, for her endless love, encouragement, and support throughout all these years. Your always positive outlook on life is truly inspiring, and my greatest source of motivation.

Sydney, Australia   Dominic Rüfenacht
August 2016

Contents

Acronyms and Notations

Acronyms

BAM	Base-anchored motion
BIHA	Bidirectional hierarchical anchoring
BMC	Block motion compensation
BOA-TFI	Bidirectional, occlusion-aware TFI
CAW	Cellular affine warping
DCT	Discrete cosine transform
DFLM	Disocclusion and folding likelihood map
DFT	Discrete Fourier transform
DWT	Discrete wavelet transform
EBCOT	Embedded block coding with optimized truncation
FOA-TFI	Forward-only anchored motion TFI
FOHA	Forward-only hierarchical anchoring
FRUC	Frame rate up-conversion
GOP	Group of pictures
HEVC	High-efficiency video coding
HST-BPI	Hierarchical spatiotemporal breakpoint induction
HVS	Human visual system
JPIP	JPEG2000 interactive protocol
MCP	Motion-compensated prediction
MCTF	Motion-compensated temporal filtering
MF	Motion field
MSE	Mean squared error
OBMC	Overlapped block motion compensation
OBME	Overlapped block motion estimation
PCRD	Post compression rate–distortion
PSNR	Peak signal-to-noise-ratio
R–D	Rate–distortion
ROI	Region of interest
SAD	Sum of absolute differences

SHVC	Scalable high-efficiency video coding
SVC	Scalable video coding
SWCA	Selective wavelet coefficient attenuation
TFI	Temporal frame interpolation
TID	Triangle identifier
TME	True motion estimation
TV	Total variation
WSVC	Wavelet-based scalable video coding

Notations

f_i — Frame at "time" instance i.

F_i — Foreground triangle ID map at frame f_i.

$\mathcal{B}_i$ — Base mesh, anchored at frame f_i.

$\mathcal{M}_i$ — Mapped mesh, anchored at frame f_i.

V_i^k — Vertex k of a triangular mesh, anchored at frame f_i.

$M_{i \rightarrow j}$ — Motion field anchored at frame f_i, pointing to frame f_j. Each *integer* location $\mathbf{m}$ in f_i has a motion vector assigned that links it with frame f_j.

$T_{i \rightarrow j}$ — Affine interpolated motion field, which maps each *continuous* location X_i in frame f_i to location X_j in frame f_j .

$\hat{D}_{i \rightarrow j}$ — Disocclusion and folding likelihood map anchored at frame f_i, estimated on motion field $M_{i \rightarrow j}$.

$\mathbf{u}$ — Motion vector $\mathbf{u} = (u, v)$, where u and v denote the horizontal and vertical component, respectively.

$\hat{S}_{i \rightarrow j}$ — Disocclusion mask anchored at frame f_i; its values are nonzero only at locations that are visible in frame f_j.

List of Figures

List of Tables

Chapter 1
Introduction

Video content accounted for almost two thirds of the world's consumer internet traffic in 2014, and is predicted to account for 80% by 2020; by the end of this decade, it is expected that almost *one million minutes* of video content will cross global IP networks *every second* [1]. According to Sandvine's 2016 "Global Internet Phenomena Report" [2], *video streaming* accounts for over 60% of peak-hour broadband internet traffic consumption in North America, with Netflix (35%) and YouTube (18%) being the main contributors. In a video streaming scenario, a variety of users with different resources in terms of screen size, resolution, processing power, and network bandwidth, are accessing the same video content, as illustrated in Fig. 1.1. Currently, the heterogeneous requirements of web streaming are met by storing *hundreds* of copies of the same video on the server [3]. Clearly, there exists a lot of redundancy between the different copies; the reason for this "wasteful" storage is that existing video coding standards (e.g., H.264/AVC [4] and HEVC [5]) are optimized for a predefined set of network and decoder constraints.

Scalable video compression presents a promising solution to the above-mentioned problem. Instead of coding the same video at different quality levels, the media is encoded in an *embedded* way, such that partial bit-streams can be decoded at lower resolution (spatial scalability), frame-rate (temporal scalability), as well as quality (SNR scalability). The importance of scalability for the delivery of video content over heterogeneous networks is evidenced by the fact that the two latest video compression standards both have scalable extensions (H.264/SVC [6] and SHVC [7]). However, being extensions of single-layer codecs, they inherit the predictive feedback loop that is used to exploit temporal redundancy between frames, which severely hampers scalability; in fact, for practical applications, the number of different quality layers is limited to just a few.

The last decade has witnessed a trend towards higher resolutions (ultra-high definition 4 and 8 K formats) and higher framerates in video. Latest mobile phones offer video recording capabilities of up to 4 K (2160 p) at 30 frames per second, making

© Springer Nature Singapore Pte Ltd. 2018
D. Rüfenacht, *Novel Motion Anchoring Strategies for Wavelet-Based Highly Scalable Video Compression*, Springer Theses,
https://doi.org/10.1007/978-981-10-8225-2_1

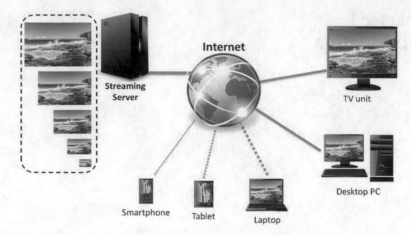

Fig. 1.1 In a video streaming scenario, multiple users are accessing the same video over a hetero-geneous network. Their devices also have quite different display sizes, resolutions, and bandwidth. Existing compression systems are ill-suited to serve the large variability, and video content has to be encoded multiple times

the *creation* of high (to ultra high) resolution video content ubiquitous. However, 4 K displays are still relatively uncommon, both in TV units as well as mobile phones, which means that in most cases, there are more pixels recorded than can be displayed. Recently, cameras that record full 360° video content emerged [8], where the user can freely choose which portion to view. With current video compression standards, the full 360 video has to be streamed, even though only a small portion is actually displayed. These are just some examples that indicate that "efficient" *interactive browsing of video* content will become increasingly important over the next decade, which requires some fundamental changes in the way videos are encoded. More "feature-rich" coders that offer high scalability and *region of interest (ROI)* access to decode only a "window" of the actual encoded content would be highly beneficial.

JPEG2000 [9] is a family of standards that provides high scalability and accessi-bility features for image (and video) content. *JPEG2000 interactive protocol (JPIP)* [10] brings an interactive browsing framework to JPEG2000, which provides very interesting opportunities for interactive browsing of video content; however, it is currently limited to *independently* coded JPEG2000 frames, without exploiting tem-poral redundancies (known as "inter" prediction) between the frames, which lowers the compression performance.

Various *wavelet-based scalable video coding (WSVC)* schemes have been pro-posed, which extend the *discrete wavelet transform (DWT)* employed by JPEG2000 to the temporal domain; the absence of a prediction feedback loop makes such schemes much better suited for high scalability. The success of WSVC schemes has mostly been limited by the fact that scalable video coders have problems at *disconti-nuities* in the motion field [11]; band-limited sampling of motion fields smooths out the sharp transitions at moving object boundaries, creating "non-physical" motion – a value interpolated between the motion of the foreground and the background object is in general a poor predictor. A number of works in video compression have demonstrated the benefits of partitioning motion fields based on their motion discontinuities [12–14]. Recently, a *highly scalable* way of representing *discontinuities* in a progressively refineable way has been proposed [15]. So-called "breakpoints" are used to signal discontinuities to adapt the wavelet bases from crossing discontinuity boundaries, which enables an efficient, scalable coding of motion fields.

The aim of this thesis is to improve the temporal transform used in wavelet-based scalable video coding. Our work is focused more towards "physical"[1] motion fields, and breakpoints are used to explicitly and efficiently describe motion discontinuities. We show that such an approach can provide many advantages for video compression, especially for scalable video coders. The objective here is to eliminate artificial block boundaries, while efficiently describing true discontinuities in the motion flow. While the use of optical flow motion for video compression recently has gained more interest [16, 17], the motion information has so far been attached to the target frame that is to be predicted. In this thesis, we show that the combination of physical motion and motion discontinuities allows us to "flip" the anchoring of motion fields to reference frames compared to conventional video codecs, which turns out to have a number of advantages; for example, foreground/background objects can be identified which are required to perform motion inference across time.

In such a reference-based motion field anchoring, the motion-compensation prediction found in traditional compression schemes is essentially replaced by *temporal frame interpolation (TFI)*. That is, in order to serve as prediction reference, motion information has to be mapped from reference to the target frame. We show in this thesis how motion discontinuities can be used to handle traditionally problematic regions around moving objects. During the motion warping process, disoccluded regions are readily observed. Because disoccluded regions are by definition regions in the target frame that are *not visible* from the respective reference frame, this information is highly valuable to guide the bidirectional prediction process of the target frames; in traditional video codecs, disocclusion information has to be *explicitly* communicated as side-information. While the main goal of this thesis is to improve the temporal scalability of video coders, because of its intrinsic connection with TFI, we also make contributions to the field of TFI. Throughout the thesis, we use TFI as a convenient way of assessing the impact of suggested improvements, without the need to build a full compression system.

We investigate three reference-based motion anchoring strategies, with progressively simpler motion anchorings. In the *bidirectional hierarchical anchoring (BIHA)* framework, the anchoring of *all* motion fields is flipped with respect to the popular hierarchical B-frame structure employed in existing video codecs. We propose an analytical model in order to determine weights for the spatio-temporal subbands of texture, motion, and breakpoint information in the BIHA scheme. The temporal hierarchy enables the prediction of finer level motion information from coarser levels in

---

[1]Throughout the thesis, we use the term "physical" motion to refer to the *apparent* physical motion of rigid objects, i.e., the projection of 3D moving objects onto the 2D camera sensor plane.

the hierarchy, which reduces the coding cost of motion fields compared to the traditional anchoring of motion at target frames. In the second motion anchoring strategy, all coded motion information is anchored so that it points forward in time; we refer to this anchoring as *forward-only hierarchical anchoring (FOHA)*. The FOHA scheme is not only conceptually simpler, but has an important positive impact on the quality of the interpolated frames.

The favourable properties of the BIHA and FOHA schemes, which both employ a *hierarchical* motion anchoring, led us start the exploration of a third motion anchoring strategy, motivated primarily by the obstacles encountered to efficiently compose motion fields across the temporal hierarchy. To this end, we propose the *base-anchored motion (BAM)* scheme, where all coded motion information is anchored at the first frame of a *group of pictures (GOP)*. This anchoring has a number of additional benefits with respect to the other two motion anchoring schemes proposed in this thesis. In particular, it facilitates the incorporation of higher-order motion descriptions, which improve the *prediction* performance for objects that are following *non-constant* velocity trajectory, which is essential for good compression performance as it reduces the prediction residual that needs to be coded. Furthermore, while not explored in this thesis, the fact that all motion is centrally organized can be expected to greatly facilitate ROI access – a feature that becomes more important as the spatial and temporal resolution of video content continues to increase.

The rest of this thesis is organized as follows: In Chaps. 2 and 3, we present important concepts and literature that is relevant for the rest of this thesis. Chapters 4, 5, 6 and 7 present the contributions of this thesis; in Chap. 4, we introduce some of the main ideas and concepts that will be used in Chaps. 5, 6 and 7, where three different motion anchoring strategies for scalable video compression are investigated. In the last chapter, we conclude the thesis and discuss interesting venues for future research.

Chapter 2: This chapter forms the first of two background chapters. We present relevant concepts for image and video compression, with a focus on scalable video compression.

Chapter 3: The three motion anchoring strategies we investigate in this thesis are all *reference-frame* anchored. In order to serve as prediction references, the motion fields need to be warped from reference to target frames. As such, the fundamental building block of all three motion anchoring strategies proposed in this thesis performs TFI. In the first part of this chapter, we give an overview of common true motion estimation schemes, as well as optical flow schemes. In the second part, we review existing TFI schemes.

Chapter 4: This chapter presents the fundamental motion field operations that are used to map motion from reference to target frames in all three motion anchoring strategies. Common to all three schemes is that they employ piecewise-smooth motion fields with sharp discontinuities at moving object boundaries. We present two motion field operations, namely motion *inversion* and motion *inference*, which are used to form motion information at the target frame in order to enable a bidirectional, occlusion-aware interpolation of the target frame. Key in resolving double mappings and assigning "physical" motion in disoccluding regions (i.e., holes in the target frame) is the insight that motion discontinuities displace with the foreground object.

The two motion field operations are evaluated in a TFI scenario, where the proposed *bidirectional, occlusion-aware TFI (BOA-TFI)* method compares favourably with state-of-the-art TFI methods.

Chapter 5: In this chapter, we propose a BIHA of motion for highly scalable video compression, where the motion field anchoring is *flipped* compared to the traditional motion anchoring at target frames that is employed in state-of-the-art video codecs. At each level of the temporal hierarchy, the scheme uses BOA-TFI to create predictions of the target frames. We derive an analytical model which gives insight into how the importance of texture and motion data changes across temporal scales of the proposed spatio-temporal transform. We use a highly scalable motion discontinuity representation using "breakpoints", presented in [15]. In this chapter, we augment the breakpoint induction policies to the temporal domain, and propose a *hierarchical spatio-temporal breakpoint induction (HST-BPI)*, which makes it possible to further reduce the coding cost of breakpoints. We compare BIHA with a scalable video coder that employs a traditional anchoring, as well as with SHVC, the latest scalable video compression standard.

Chapter 6: This chapter focuses on further improving the TFI performance, which directly impacts the compression performance. We propose various changes to the BIHA scheme in order to address shortcomings that were identified in the two previous chapters. First, we propose a more robust motion discontinuity measure that is based on motion field divergence. Further, we change the direction of the motion *inference* operation to a *forward* motion inference; we call the resulting scheme *forward-only anchored motion TFI (FOA-TFI)*. As is turns out, this change leads to less ghosting in the interpolated frames. Finally, we propose two texture optimization procedures that further improve the visual (and objective) quality of the interpolated frames. The scheme is extensively evaluated in a TFI scenario. At the end of the chapter, we outline a hierarchical motion anchoring strategy for WSVC that uses FOA-TFI as building block, which we call FOHA.

Chapter 7: Both the BIHA and the FOHA scheme presented in Chaps. 5 and 6, respectively, are reference-anchored *hierarchical* schemes. In this chapter, we propose to further simplify the motion anchoring to a *base-anchored motion (BAM)*, where all *coded* motion information is anchored and organized at the first frame of a GOP. We present a mesh sparsification algorithm, which makes it possible to reduce the number of motion vectors that have to be coded, with almost no impact on the texture prediction quality. Furthermore, the base-anchoring facilitates the incorporation of acceleration (and higher order motion models), which improves the *prediction quality* in scenes that do not follow the constant velocity assumption, which is essential for improving the compression performance. We incorporate the BAM scheme into HEVC, where preliminary coding results prove illuminating.

Chapter 8: This chapter summarizes the main results of this thesis. Furthermore, we discuss a number of promising venues for highly scalable video compression that are opened up by the ideas presented in this thesis.

References

1. Cisco, Cisco visual networking index: forecast and methodology, 2015–2020, Technical Report (2016)
2. Sandvine, Global internet phenomena (2016), https://www.sandvine.com/trends/global-internet-phenomena/. Accessed 30 August 2016
3. Gigaom, To stream everywhere, netflix encodes each Movie 120 times (2012), https://gigaom.com/2012/12/18/netflixencoding/. Accessed 30 August 2016
4. T. Wiegand, G.J. Sullivan, G. Bjøntegaard, A. Luthra, Overview of the H.264/AVC video coding standard. IEEE Trans. Circuit Syst. Video Technol. **13**(7), 560–576 (2003)
5. G.J. Sullivan, J.-R. Ohm, W.-J. Han, T. Wiegand, Overview of the high efficiency video coding (HEVC) standard. IEEE Trans. Circuit Syst. Video Technol. **22**(12), 1649–1668 (2012)
6. H. Schwarz, D. Marpe, T. Wiegand, Overview of the scalable video coding extension of the H.264/AVC standard. IEEE Trans. Circuit Syst. Video Technol. **17**(9), 1103–1120 (2007)
7. P. Helle, H. Lakshman, M. Siekmann, J. Stegemann, T. Hinz, H. Schwarz, D. Marpe, T. Wiegand, A scalable video coding extension of HEVC, in *Proceedings IEEE Data Compression Conference* (2013)
8. IM360, Creating immersive experiences (2016), http://www.im360.info/. Accessed 30 August 2016
9. D. Taubman, M.W. Marcellin, *JPEG2000: Image Compression Fundamentals, Standards, and Practice* (Kluwer Academic Publishers, Boston, 2002)
10. D. Taubman, R. Prandolini, Architecture, philosophy, and performance of JPIP: internet protocol standard for JPEG2000, in *Proeedings SPIE*, vol. 5150 (2003), pp. 791–805
11. N. Adami, A. Signoroni, R. Leonardi, State-of-the-art and trends in scalable video compression with wavelet-based approaches. IEEE Trans. Circuit Syst. Video Technol. **17**(9), 1238–1255 (2007)
12. S. Milani, G. Calvagno, Segmentation-based motion compensation for enhanced video coding, in *Proceedings IEEE International Conference Image Procesing* (2011), pp. 1649–1652
13. R. Mathew, D.S. Taubman, Scalable modeling of motion and boundary geometry with quad-tree node merging. IEEE Trans. Circuit Syst. Video Technol. **21**(2), 178–192 (2011)
14. I. Daribo, D. Florencio, G. Cheung, Arbitrarily shaped sub-block motion prediction in texture map compression using depth information. Pict. Codin. Symp. 121–124 (2012)
15. R. Mathew, D. Taubman, P. Zanuttigh, Scalable coding of depth maps with R-D optimized embedding. IEEE Trans. Image Proc. **22**(5), 1982–1995 (2013)
16. G. Ottaviano, P. Kohli, Compressible motion fields, in *Proceeding IEEE Conference Computer Vision and Pattern Recognition* (2013)
17. A. Zheng, Y. Yuan, H. Zhang, H. Yang, P. Wan, O. Au, Motion vector fields based video coding, in *Proceeding IEEE International Conference Image Processing* (2015), pp. 2095–2099

Chapter 2
Scalable Image and Video Compression

This chapter introduces the fundamental concepts employed in (scalable) image and video compression schemes, which are required to understand and appreciate the contributions of this thesis. The main difference between image and video compression is that the latter can exploit both spatial *and* temporal redundancies in the data, which results in higher compression ratios than in image compression systems.

After a brief generic presentation of the main building blocks of a lossy feedforward compression system, we use JPEG2000 to give a concrete example of a state-of-the-art image compression system in Sect. 2.1. The reason we chose JPEG2000 is that it allows us to simultaneously present the fundamental concepts of a compression system, and to introduce additional *features*, e.g., scalability and accessibility. As mentioned in the introduction, this thesis investigates new motion anchoring strategies with the aim of bringing the scalability and accessibility features of JPEG2000 to video compression systems, in order to facilitate interactive browsing of video.

In Sect. 2.2, we introduce the *hybrid* compression scheme, which is used by all existing standardised video codecs; a particular focus will be on the motion-compensated prediction feedback loop. As we will see in Sect. 2.3, this feedback loop imposes restrictions on the scalability attributes of hybrid coding schemes. We then shift the focus to wavelet-based scalable video compression schemes, which have an open-loop structure that makes them better suited for highly scalable video compression. In this thesis, we focus on physical motion rather than block motion, which is characterized as being piecewise-smooth with discontinuities around moving objects. At the end of this chapter, we present a relatively recent way of compressing such motion fields using a motion discontinuity representation that is highly scalable both in resolution and precision.

© Springer Nature Singapore Pte Ltd. 2018
D. Rüfenacht, *Novel Motion Anchoring Strategies for Wavelet-Based Highly Scalable Video Compression*, Springer Theses,
https://doi.org/10.1007/978-981-10-8225-2_2

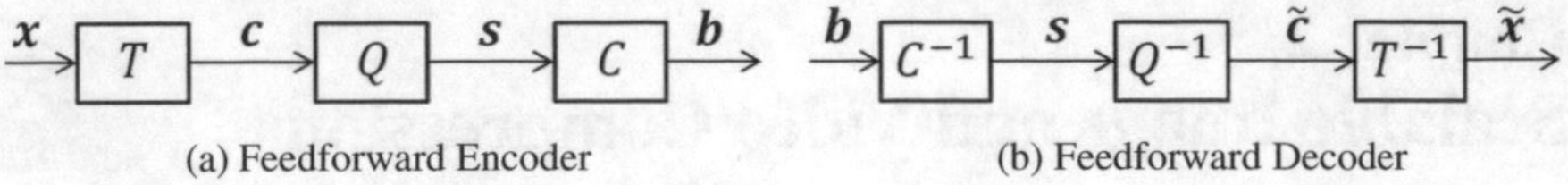

$$x \rightarrow \boxed{T} \xrightarrow{c} \boxed{Q} \xrightarrow{s} \boxed{C} \xrightarrow{b} \qquad \xrightarrow{b} \boxed{C^{-1}} \xrightarrow{s} \boxed{Q^{-1}} \xrightarrow{\hat{c}} \boxed{T^{-1}} \xrightarrow{\tilde{x}}$$

(a) Feedforward Encoder   (b) Feedforward Decoder

Fig. 2.1 Main blocks of a "lossy" feedforward compression system. T is the transformation, Q the quantization, and C is the coding stage. The "lossiness" is introduced by the quantizer Q

We now briefly summarize the main functions of the three main building blocks (see Fig. 2.1) of *lossy feedforward* image and video compression systems. The goal of any lossy compression system is to minimize the number of bits required to achieve a certain level of distortion; alternatively, a lossy scheme aims at minimizing the distortion for a given bit-rate budget.

Transformation The aim of the transformation stage is to convert the input data into a new representation that is better suited for the subsequent quantization and coding stages. For image and video compression applications, the input signal is typically partitioned into a collection of frequency bands; each such frequency band has different statistical properties, and the subsequent quantization and coding stages can be tailored accordingly. Furthermore, properties of the *human visual system (HVS)*, such as the fact that the HVS is more sensitive to changes in the low-frequency bands, can be exploited.

Quantization The aim of the quantization stage is to introduce controlled loss in order to achieve compression. In the interest of conciseness, we only present the main concepts of this stage, and refer the interested reader to the comprehensive overview by Gray and Neuhoff [1].

The quantizer $Q(\cdot)$ divides the region of support into N *disjoint* intervals I_j, and maps coefficients to symbols s_j, according to the interval they fall into:

$$Q(c) = s_j \text{ if } c \in I_j. \tag{2.1}$$

The reconstruction of an approximate value $\hat{c}_j$ from the corresponding symbol s_j is referred to as "dequantization":

$$Q^{-1}(s_j) = \hat{c}_j. \tag{2.2}$$

Quantization is the only significant source of distortion in an image and video compression scheme. A minimum mean-squared error dequantization strategy is to use the centroid of the relevant quantization interval as the representation for the corresponding quantization index. Increasing the number of intervals or quantization bins leads to a more accurate reconstruction of the original input, at the expense of a higher coding cost.

Coding The primary aim of the coding stage is to represent the sequence of symbols obtained in the quantization stage with the least number of bits, by exploiting the statistical redundancy between different symbols. For typical image and video content, not all symbols are equally probable. By assigning shorter codewords to symbols that are more probable, a shorter overall bit-stream can be achieved; this is referred to as "entropy coding". In the literature, such schemes are known as "variable-length coding" schemes. Huffman coding [2] and arithmetic coding [3] are popular examples of variable-length coding schemes. In Huffman coding, the length of a codeword is proportional to the amount of information of the respective symbol. In arithmetic coding, variable-length codes are assigned to variable-length blocks of symbols; in theory, an arithmetic coder can achieve an average bit-rate that is arbitrarily close to the entropy of the source. In addition, and perhaps more importantly, incremental encoding and decoding can be realized, which enables the realization of long, efficient codes without requiring a large amount of memory to maintain a large collection of codewords.

At the decoder, all the operations are inverted and applied in reverse order (see Fig. 2.1b). While these three fundamental stages have been introduced as separate building blocks, it is important to note that they go hand in hand. In the next section, we show how these main building blocks are carefully designed to work together, on the specific example of JPEG2000.

2.1 JPEG2000: Scalability, Accessibility, and Intrinsic Upsampling

In this section, we present the main methods and techniques employed by JPEG2000 [4] that enable its high scalability, accessibility, and "intrinsic upsampling"[1] features. The latter can be seen as part of a highly scalable compression system. However, as we will see in Sect. 2.3, intrinsic upsampling is not easily possible in the temporal domain. In fact, a *separate* field of research dedicated to increasing the framerate of video at the decoder exists, which is usually referred to as TFI, or framerate upsampling. The relevant research of this separate field of research will be reviewed in the next chapter. As we will show in this thesis, the intrinsic temporal upsampling property of the proposed highly scalable compression schemes achieves highly competitive TFI results.

In image compression, *scalability* refers to an encoding of the image in an *embedded* way, such that lower resolutions are embedded in higher resolutions, and lower qualities are embedded in higher qualities; importantly, there are *multiple dimensions of embedding*. A scalable encoding of an image can be decoded at various qualities and resolutions.

---

[1] "Intrinsic upsampling" is a term we use to emphasize that in JPEG2000, the decoder can decide to decode the image at a higher resolution than what was available at the encoder; this is a direct consequence of the subband transform that is employed, and not normally mentioned as a separate feature. Admittedly, in the spatial domain, similar results can be achieved using other well-known techniques, such as bilinear, bi-cubic, or sinc-interpolation.

Accessibility refers to the ease with which a user can select an ROI in the image that will be decoded at higher quality than the rest of the image. Accessibility requires that the codestream is organized in a way that limits the amount of data that need to be decoded that are not part of the ROI. This is particularly appealing for interactive browsing of media, as mentioned in the introduction.

We now present the main building blocks of JPEG2000 that enable state-of-the-art compression performance with a number of attractive features that go beyond compression.

2.1.1 Subband Transform for Resolution Scalability

In contrast to block-transforms such as the well-known *discrete cosine transform (DCT)* [5], subband transforms have overlapping regions of support for the analysis and synthesis operators, and hence they do not suffer from blocking artefacts, and are able to better exploit statistical redundancies in the input data.

The transformation from the input signal to the different subbands is called *analysis*, and the reverse transformation, which takes the individual subbands and reconstructs the signal $\tilde{x}[m]$, is referred to as *synthesis*. The analysis stage consists of applying a set of bandpass filters to the input signal, each extracting a different frequency band. Since these analysis filters are not ideal bandpass filters, aliasing is introduced into the subbands. Careful design of the corresponding synthesis filters makes it possible to cancel out the aliased parts, such that the reconstructed signal $\tilde{x}[m]$ will be identical to the input signal $x[m]$; classical results for the design of aliasing-free, linear-phase analysis/synthesis filters are described in [6, 7].

For image and video compression, cascaded *dyadic* filter banks are most widely used. In this case, the input signal $x[m]$ is filtered by low-pass and high-pass filters h_0 and h_1, respectively, and then subsampled by a factor of 2. The output of the analysis stage are two subbands y_0 and y_1, which contain the low and high frequency parts of the image, respectively. The synthesis part consists of reversing the analysis part. That is, the signal is first upsampled by inserting a zero between every pair of samples $y_j[m]$, for $j \in \{0, 1\}$, followed by filtering the upsampled signal with g_j.

Since most information of natural images is contained in the low-frequency subband, so-called *tree-structured* subband transforms have been adopted, where at each decomposition level, only the low-frequency subband is further decomposed; Fig. 2.2 depicts the 1D-case of such a *dyadic tree-structured filter bank* [6].

The 2D *discrete wavelet transform (DWT)* filters are obtained through *separable extension*; that is, the 2D filtered output $y[\mathbf{m}] = y[m, n]$ is obtained by successively applying the 1D-DWT along the rows and columns of the input image $x[\mathbf{m}] = x[m, n]$. The spatial subbands can then be obtained by de-interleaving $y[\mathbf{m}]$:

$$y_{i,j}[m, n] = y[2m + i, 2n + j], \text{ for } i, j \in \{0, 1\}. \tag{2.3}$$

In order to emphasize the orientation of the subbands, they are commonly labelled as follows: $y_{0,0}$ as LL, $y_{0,1}$ as HL, $y_{1,0}$ as LH, and $y_{1,1}$ as HH, where the first

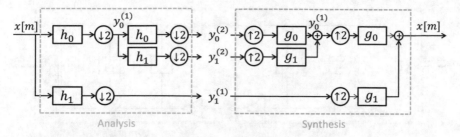

Fig. 2.2 2-level dyadic tree filter bank in one dimension

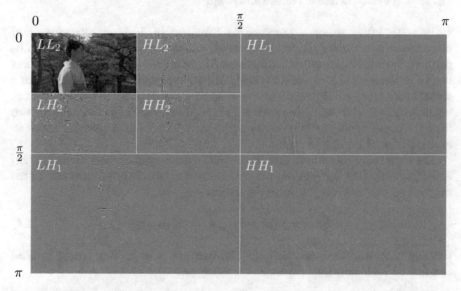

Fig. 2.3 Example of a 2D subband decomposition. For visualization, all but the LL_2 subband have been normalized and offset, so that gray signifies a value of zero. One can see how the main energy is contained in the LL_2 subband

and second letter stand for the type (e.g., low-pass L or high-pass H) of the applied horizontal and vertical filter, respectively. For example, HL is the signal obtained by applying a horizontal high-pass filter, and a vertical low-pass filter; hence, the HL band represents the vertical edges. In a tree-structure, the subbands are subscripted with the corresponding level in the hierarchy. Figure 2.3 shows an example of a dyadic 2D-subband decomposition with two decomposition levels.

The LL_2 subband is a spatially subsampled version of the original image. The other subbands contain high-frequency content needed to reconstruct the image at a higher spatial resolution. For example, all information needed to reconstruct LL_1 is contained in LL_2, together with HL_2, LH_2, and HH_2. This inherent *resolution-scalability* makes subband transforms particularly appealing for scalable compression schemes.

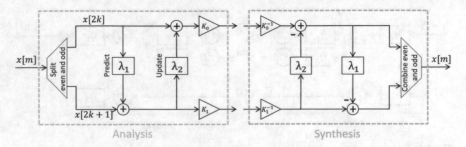

Fig. 2.4 Lifting implementation with two lifting steps

Lifting The lifting structure proposed by Sweldens [8, 9] is an alternative way of constructing a multiresolution representation of a signal, which turns out to be particularly useful for this thesis. All symmetric *finite impulse response (FIR)* filters can be implemented by successfully updating the even and odd subsequences of a signal by a sequence of lifting steps. As we shall see in Sect. 2.3.3.3, filters that can be implemented with two lifting steps are of particular interest; in this case, the two lifting steps are usually referred to as *predict* and *update* steps. Figure 2.4 shows an example of such a lifting structure.

The signal $x[m]$ is split up into its odd and even polyphase components, $x[2k]$ and $x[2k + 1]$. Odd samples are predicted from the adjacent even samples in the neighbourhood $\mathcal{N}$, which produces the residue:

$$h[m] = x[2k + 1] - P\{(x[2k])_{k \in \mathcal{N}}\} \tag{2.4}$$

Because $x[2k]$ potentially contains a lot of aliased components, it should be updated to a filtered version $l[k]$, which is done in the *update* step:

$$l[k] = x[2k] + U\{(h[k])_{k \in \mathcal{N}}\} \tag{2.5}$$

The signal x can be perfectly reconstructed by undoing the update and predict steps as follows:

$$x[2k] = l[k] - U\{(h[k])_{k \in \mathcal{N}}\}$$
$$x[2k + 1] = h[k] + P\{(x[2k])_{k \in \mathcal{N}}\} \tag{2.6}$$

Of particular interest for this thesis is the 5/3 biorthogonal filter [10], which can be implemented using the following *two* lifting steps:

$$\lambda_1(z) = -\frac{1}{2}(1 + z)$$
$$\lambda_2(z) = \frac{1}{4}(1 + z^{-1}), \tag{2.7}$$

and the *gain factors* $K_0 = 1$ and $K_1 = \frac{1}{2}$. With this filter, the predict and update step become:

$$h[k] = x[2k+1] - \frac{1}{2}\left(x[2k] + x[2k+2]\right)$$

$$l[k] = x[2k] + \frac{1}{4}\left(x[2k-1] + x[2k+1]\right). \tag{2.8}$$

One can see how each odd pixel is predicted from the neighbouring even pixels. The predicted value is then subtracted from the odd pixel, which will only contain the part that could not be predicted (i.e., the high-frequency part). After all odd pixels have been predicted, some of the prediction error of the predicted odd pixels is fed back to the even pixels during the update step. The importance of this update step is that during synthesis, it distributes the high-pass quantization error across space (or time), while shaping its spectrum, so that the overall distortion is reduced.

While this section focused on image transforms, we will see in Sect. 2.3.2 how subband transforms can be applied in the temporal domain, where inherent multiresolution representation naturally lends itself to highly scalable video compression schemes.

2.1.2 Deadzone Scalar Quantization

As mentioned earlier in this section, the aim of the quantization stage is to convert continuously valued transform coefficients c_j to a finite set of symbols $\{s_j\}$. JPEG2000 employs a deadzone scalar quantizer, which can be seen as a scalar quantizer where the center quantization interval is wider than the others; the width of the center interval I_0 is set by ϵ. Denoting Δ as the quantization step size, the deadzone quantizer assigns symbols according to the following equation:

$$s_j = \begin{cases} \text{sgn}(c_j) \cdot \left\lfloor \dfrac{|c_j|}{\Delta} + \epsilon \right\rfloor & \dfrac{|c_j|}{\Delta} + \epsilon > 0 \\ 0 & otherwise \end{cases}, \tag{2.9}$$

where $\lfloor \cdot \rfloor$ is the floor function (round down to the nearest integer). In JPEG2000, the width of I_0 is set as 2Δ, which is obtained by setting $\epsilon = 0$.

The dequantizer which maps symbols s_j back to (approximated) coefficient values $\hat{c}_j$, is

$$\hat{c}_j = \begin{cases} \text{sgn}(s_j) \cdot (|s_j| - \epsilon + 0.5) \cdot \Delta & s_j \neq 0 \\ 0 & otherwise \end{cases}. \tag{2.10}$$

2.1.3 Embedded Coding for Distortion Scalability

An embedded coder creates a bit-stream that has many identifiable parts, which makes it possible that the video can be decoded at a number of spatial resolutions, frame-rates, and qualities. The aim is that the *rate-distortion (RD)* performance of each decoded sub-stream is comparable to a corresponding single layer coding at the same spatio-temporal resolution and quality. Examples of achieving embedded image compression include the well-known embedded zero-tree wavelet (EZW) [11] and set partitioning in hierarchical trees (SPIHT) [12], as well as *embedded block coding with optimized truncation (EBCOT)* [13], which is employed in JPEG2000.

2.1.3.1   Embedded Block Coding with Optimized Truncation (EBCOT)

In EBCOT, the subbands obtained after wavelet analysis are partitioned into smaller blocks called "code-blocks", typically of size 64×64 or 32×32 [4]. The samples of each such code-block are *independently* coded using a bit plane coding process, which is referred to as "tier-1" coding.

Starting with the most significant bit plane, EBCOT applies three coding passes to each bit plane: (1) Significance propagation pass (SPP); (2) magnitude refinement pass (MRP); and (3) cleanup pass (CP). With the exception of the most significant bit plane, where only the cleanup pass is performed, the significance pass is the first coding pass performed. We now provide more details about the three coding passes.

In the SPP, each bit plane is associated with a significance threshold, which is a power of two. The initial threshold is determined by the maximum magnitude of all the wavelet coefficients, and divided by two after the coding of each bit plane. The significance of each sample that is predicted to be significant from its neighbours is represented with one binary symbol. If the sample is found to be significant, its sign is represented with one binary symbol. The symbols are then encoded using context-adaptive arithmetic coding that exploits statistical redundancies. Every sample that was found to be significant in a previous bit plane is refined at each lower bit plane in the refinement pass, represented using one binary "refinement" symbol, during the MRP. The CP is conceptually similar to the significance pass, except that it deals with samples that have not yet been found to be significant and that are predicted to be insignificant for that coding pass.

Each coding pass at each bit plane creates a sequence of symbols, and context modelling is employed to exploit higher-order statistical redundancies that exist between bit planes. The main idea behind context-adaptive modelling is that if a not yet significant sample is surrounded by samples that have become significant in an earlier bit plane, it has an increased probability of becoming significant.

Post-Compression Rate-Distortion Optimization The selection of truncation points can be deferred until the code-blocks have been independently coded, when the rates and associated distortions will be known. The coding information generated in the tier-1 is grouped together into *packets*. A subset of these packets is then selected in an R-D optimization algorithm to minimize the distortion for a given bit-rate. They are grouped together into so-called "quality layers", which enable progressive refinement of the quality of the video.

2.2 Hybrid Video Compression

In this section, we present the "hybrid" video compression scheme, which is employed by all standardised state-of-the-art video codecs; Fig. 2.5 shows a (simplified) block diagram of a hybrid codec. For conciseness, we will not delve into all the intricate details that are employed in different standardised codecs; for a comprehensive overview of the two latest standardised video codecs, the interested reader is referred to [14] for H.264/AVC, and [15] for HEVC, respectively. The purpose of this section is to present the fundamental structure of a hybrid video codec, which will be useful for the discussion about scalable video coding in Sect. 2.3. The scheme is called a *hybrid* since it combines *transform* coding in the spatial domain (see Fig. 2.5a) with a *predictive feedback loop* to perform motion-compensated prediction in the temporal domain [16]. The difference between transform and predictive coding schemes is that the former is feedforward, whereas predictive techniques involve a feedback loop. As we shall see, this predictive feedback loop is one of the main obstacles for efficient scalable video compression in standardised video codecs. In Sect. 2.3.2, we present how open-loop video compression systems (i.e., transform coding only) are much better suited for high scalability.

In hybrid coding systems, certain "key" frames are coded with spatial information from the same frame only; these are referred to as $INTRA$ coded frames. Modern video codecs use directional spatial prediction, which by itself is a kind of feedback loop, but one which does not operate in time. In the next section, we show how the temporal predictive feedback loop is used in hybrid compression schemes to exploit temporal redundancies; this so-called $INTER$ mode results in a significant improvement of the compression ratio.

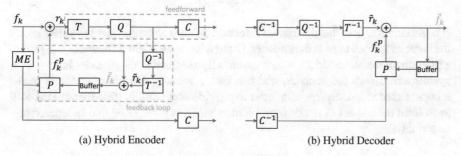

(a) Hybrid Encoder   (b) Hybrid Decoder

Fig. 2.5 Main building blocks of a hybrid codec, as used by all standardised video codecs. Note the predictive feedback loop. Only the prediction residuals and side-information (e.g., motion vectors, prediction mode, ...) are transmitted, which can result in big compression improvements

2.2.1   Motion-Compensated Prediction

At the heart of any modern video coder is *motion-compensated prediction (MCP)*, where the temporal correlation between frames is exploited to predict certain *target* frames from neighbouring *reference* frames; only the prediction residual and side information (e.g., motion vectors) are encoded, which can result in significant compression gains. A target frame can either be unidirectionally predicted from a preceding (or succeeding) reference frame, or bidirectionally predicted from both previous and succeeding reference frames. Hybrid coders implement MCP by adding a predictive feedback loop to the feedforward scheme, as shown in Fig. 2.5.

Let us focus on the most simple case where the current frame is predicted using only the preceding frame. The motion between the current frame f_k (at the discrete time instant k), and the preceding frame (f_{k-1}), is estimated in the motion estimation (ME) phase, resulting in a set of motion vectors (MV). These are used to create a motion-compensated prediction (P) of the target frame f_k; let us denote the predicted frame as f_k^p. Next, the difference between the motion-compensated prediction of the target frame and the actual frame is computed:

$$r_k = f_k - f_k^p, \tag{2.11}$$

which is commonly referred to as *prediction residual*. The residuals are then transformed and quantized, and used to predict the next frame. On the decoder side, the coding, quantization, and the transformation are inverted, which results in $\tilde{r}_k$. Then, for every $INTER$-coded frame, the necessary motion-compensated predictions f_k^p are formed, which are added to the residual $\tilde{r}_k$, to obtain the reconstructed frame $\tilde{f}^k$,

$$\tilde{f}^k = \tilde{r}_k + f_k^p. \tag{2.12}$$

It is important to note that the motion information used at the decoder has to be exactly the same as the one used at the encoder. Otherwise, even slight differences can lead to a phenomenon called "drift", which essentially means that the decoder is no longer synchronized with the encoder, and can cause visually disturbing artefacts in the reconstructed video sequence. In order to provide accessibility (fast-forward) as well as to limit the impact of the drift problem, a video is partitioned into independently coded GOPs.

2.2.1.1   Group of Pictures (GOP)

In a video coder, frames are grouped together to *group of pictures (GOP)*; Fig. 2.6 shows two commonly used GOP structures,[2] where the three most common types of frames can be identified: I, P, and B.

---

[2]Note that in most video compression literature, the arrows point to the frame that is to be predicted. In order to be consistent with the rest of this thesis, the arrows are anchored at the frame where the

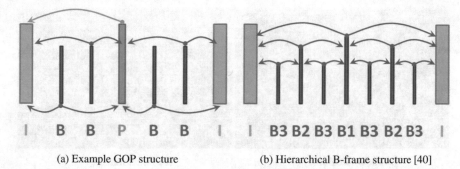

(a) Example GOP structure   (b) Hierarchical B-frame structure [40]

Fig. 2.6 Examples of GOP structures. I-frames are intra-coded frames, which are not predicted from other frames; P-frames are predicted from the previous frame, and B-frames are bidirectionally predicted from previous and succeeding reference frames (I or P). **a** Shows a common structure which involves I, P, and B-frames; **b** shows a so-called "hierarchical B-frame" structure, which is of interest in this thesis, since it enables temporal scalability in hybrid compression schemes

I-frames are intra-coded frames, which means that they are not predicted from any other frames; such frames are also referred to as "key frames". Since they are not dependent on any other frames, they are useful as they "reset" any potential drift errors, and enable fast temporal access. P-frames are predicted from a *previous* frame in the GOP, which can either be an I-frame, or another (already decoded) P-frame. Lastly, B-frames are *bidirectionally* predicted from previous and future (decoded) frames. They are the most efficient in terms of reducing the prediction residual, especially in regions that are not visible (e.g., occluded) in the previous reference frame. This is because there is a good chance that regions that are occluded in a previous frame are visible in the succeeding reference frame; by sending occlusion information as *side-information* along with the bidirectional motion field, the residual around moving objects can be significantly reduced. It is worth noting that parts of a P or B-frame can be Intra-predicted, but not vice-versa; this will be further discussed in Sect. 2.2.1.3.

Figure 2.6b shows a hierarchical B-frame structure [17], where B-frames are only predicted from lower-indexed B-frames or I-frames; in this structure, the video can be decoded at different frame-rates by dropping higher-indexed B-frames. We will go into more detail of this special form of temporal scalability in Sect. 2.3, and turn our attention to the estimation of motion in hybrid video coders.

2.2.1.2 Unconstrained Block Matching

From an implementation point of view, it seems a natural choice to "anchor" the motion field at the frame that is to be predicted (e.g., the *target* frame). Almost every

---

motion is described, which means that the arrow direction is reversed to what is commonly found in the literature.

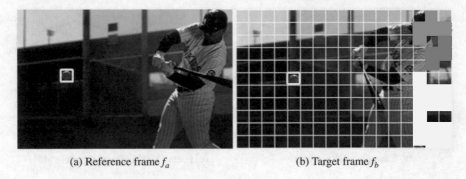

(a) Reference frame f_a (b) Target frame f_b

Fig. 2.7 Block Matching example. In the example, the target frame is partitioned into blocks. For each such block, the best match (in terms of smallest prediction residual) is found in the reference frame

successful video coder employs *block motion compensation (BMC)* [18], where the *target frame* f_b is partitioned into N disjoint blocks, and for each such block K_l, the "motion" vector $\mathbf{u}_l$ is found that best predicts the block from the reference frame f_a (see Fig. 2.7). That is,

$$\mathbf{u}_l = \arg\min_{\mathbf{u}} E_D(f_a, K_l, \mathbf{u}), \tag{2.13}$$

where E_D is the *block distortion measure (BDM)*. Using $\mathbf{u}$ to denote the displacement vector at location $\mathbf{m}$, and $\rho(\cdot, \cdot)$ as the *matching criterion*, which in most video compression schemes is the *mean squared error (MSE)*, e.g., $\rho(i, j) = (i - j)^2$, the BDM can be written as:

$$E_D(f_a, K_l, \mathbf{u}) = \sum_{\mathbf{m} \in K_l} \rho\left(f_a[\mathbf{m} + \mathbf{u}], f_b[\mathbf{m}]\right). \tag{2.14}$$

It is worth noting that the motion is normally not integer-valued, and hence sub-pixel search, achieved by upsampling the reference frame and searching on the integer grid of the upsampled frame, is employed; quarter and even eighth-pixel precision is commonly used [14, 19]. Figure 2.8 shows the impact of block size on prediction residual for a P-frame.

One can see how smaller block sizes lead to better predictions, but also result in less smooth motion fields. In the extreme case of a block size of 1×1, the best prediction in terms of smallest prediction residual can be achieved. However, in a coding environment, the cost of coding the motion has to be balanced with the cost of coding the prediction residual. An efficient video coder therefore strikes to find the optimal *balance* between the cost of coding the motion and the cost of coding the prediction residual, which is found using an R-D optimization framework.

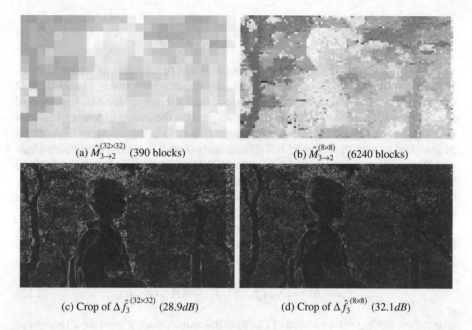

(a) $\hat{M}_{3\rightarrow 2}^{(32\times 32)}$ (390 blocks)   (b) $\hat{M}_{3\rightarrow 2}^{(8\times 8)}$ (6240 blocks)

(c) Crop of $\Delta \hat{f}_3^{(32\times 32)}$ (28.9dB)   (d) Crop of $\Delta \hat{f}_3^{(8\times 8)}$ (32.1dB)

Fig. 2.8 Impact of block size on motion-compensated prediction error. We show two (colour-coded) block motion fields obtained using unconstrained block matching; the colour-code is explained in Fig. A.1. Larger blocks result in a less noisy motion field. However, the prediction performance is significantly reduced compared to smaller prediction blocks

2.2.1.3 Rate-Distortion Optimization in Hybrid Codecs

There is no explicit attempt in standardised video coders to estimate the "true" motion of the scene. Instead, the motion is chosen in an R-D optimal way. That is, the objective is to get the minimum distortion subject to an upper bound on the overall bit-rate; or, equivalently, to get the minimum bit-rate subject to an upper bound on overall distortion. Both objectives are equivalent to minimizing

$$J = D + \lambda R, \tag{2.15}$$

where D is the overall distortion and R is the overall bit-rate, while $\lambda > 0$ determines the constraint (on distortion or bit-rate) for which the solution is optimal.

The distortion of an approximation $\hat{f}$ with respect to an original frame f is usually measured in terms of MSE:

$$MSE \triangleq \frac{1}{MN} \sum_{m=1}^{M} \sum_{n=1}^{N} \left(f[m, n] - \hat{f}[m, n] \right)^2, \tag{2.16}$$

where M and N are the width and the height of the input frame. In image and video compression, the "quality" is most commonly measured in terms of *peak signal-to-noise-ratio (PSNR)*, which is based on the MSE, and defined as

$$PSNR \triangleq 20 \log_{10} \left(\frac{2^B - 1}{\sqrt{MSE}} \right), \tag{2.17}$$

where B corresponds to the number of bits required to represent an image sample. In order to evaluate the performance of a video compression scheme, so-called "R-D curves" are created, where the PSNR is plotted over a range of bit-rates. Bjøntegaard [20] proposed a simple model to quantitatively assess the coding efficiency between two video compression algorithms. The main idea is to fit a third-order logarithmic polynomial to a set of $N \geq 4$ PSNR measurements with corresponding bit-rate values; then, an approximation of the average PSNR difference (BD-PSNR) is obtained by calculating the difference between the integrals of the fitted R-D curves, divided by the integration interval. Similarly, the so-called "BD-Rate" can be computed, which expresses the average bit-rate difference (in %) over the whole range of PSNRs.

In existing video codecs, the solution to (2.15) is found on a per-block basis. That is, for each block K_l, the *block prediction mode* I_l (see end of this section for more details) is found by minimizing the following Lagrangian cost function [21]:

$$J(K_l, I_l) = D_{rec}(f_a, K_l, I_l) + \lambda R_{rec}(K_l, I_l). \tag{2.18}$$

In practice, finding the minimum of (2.18) is infeasible, since it involves all blocks of all frames of the video sequence to be compressed. A number of simplifications aiming at reducing the computational complexity have been proposed; a good overview can be found in [22]. A widely accepted strategy is to first find the motion vectors for each block, using

$$\mathbf{u}_l = \arg \min_{\mathbf{u} \in U} E(f_a, K_l, \mathbf{u}) + \lambda_{mv} R_{mv}(K_l, \mathbf{u}), \tag{2.19}$$

where U is the set of all possible *partitions* of the block K_l that are allowed by the standard. For example, H.264/AVC [14] considers block sizes of 16×16, 8×8, and 4×4, as well as rectangular blocks of size 16×8, 8×16, 8×4, and 4×8.

We now have a closer look at three commonly used block prediction modes $I_l \in \{INTRA, INTER, SKIP\}$. $INTRA$ and $INTER$ have been introduced in Sect. 2.2; in the $SKIP$ mode, a block "inherits" the motion from its causal neighbours to form a prediction, *without* texture residual coding. The distortion for the $INTER$ mode of a block K_l is computed as follows:

$$D_{rec}(K_l, INTER, \mathbf{u}_l) = \sum_{\mathbf{m} \in K_l} (f_a[\mathbf{m} + \mathbf{u}_l] - f_b[\mathbf{m}])^2, \tag{2.20}$$

and $R_{rec}(K_l, INTER)$ is the sum of the rates for the motion vectors, transform coefficients, and mode information.

Using $\mathbf{u}_l^{SKIP}$ to denote the motion inherited from the causal neighbours of block K_l, the distortion for the $SKIP$ mode can be found as follows:

$$D_{rec}(K_l, SKIP, \mathbf{u}_l^{SKIP}) = \sum_{\mathbf{m} \in K_l} (f_a \left[\mathbf{m} + \mathbf{u}_l^{SKIP} \right] - f_b[\mathbf{m}])^2, \qquad (2.21)$$

and the rate $R_{rec}(K_l, SKIP)$ is (approximately) one bit; this means that in the $SKIP$ mode, there is no texture residual coding involved.

Lastly, the distortion for the $INTRA$ mode is:

$$D_{rec}(K_l, INTRA) = \sum_{\mathbf{m} \in K_l} (f_b[\mathbf{m}] - \hat{f}_b[\mathbf{m}])^2, \qquad (2.22)$$

and $R_{rec}(K_l, INTRA)$ is the rate obtained after entropy coding of the texture residual. From the above elaborations, one can see the *opportunistic* nature of the motion information obtained in a hybrid video coder.

2.2.1.4 Problems with Block-Motion and Ways of Addressing Them

We now state the two main issues with block-based motion estimation, and present some key approaches that have been developed to mitigate these problems:

1. Blocks are unable to represent motion in the vicinity of moving object boundaries. This means that any block that straddles a motion boundary will be unable to represent the underlying motion;
2. Block-motion schemes typically estimate *translational* motion, and are hence unable to represent *non-rigid* motion, such as rotating objects and zooming.

The first point can be mitigated by allowing varying block sizes, which is commonly referred to as *variable block size* (VBS) prediction [23]. Each such block is assigned a different motion vector. A quadtree structure can be used to structure variably sized blocks. Mathew and Taubman [24] show how the dependency between leaf-nodes of the quadtree can be exploited using a concept called leaf-merging; a similar concept is used in HEVC [15]. The large variety of block sizes improves the ability to represent motion in the vicinity of moving object boundaries, and to a certain extent *implicitly* models discontinuities in the motion field by having smaller block sizes in the vicinity of motion boundaries. For many scenes, the object motion is smooth, in which case the underlying motion field is piecewise-smooth. In HEVC, blocks can inherit the motion of their spatial neighbours, which favours smoothness within moving objects. Figure 2.9 shows example motion fields estimated using H.264/AVC and HEVC using *only INTER* mode prediction. Note how both motion fields are reasonably smooth within objects; however, the merge mode of HEVC seems to be highly effective around object boundaries, where H.264/AVC yields a lot more erroneous motion "prediction" vectors. Another way of handling blocks that straddle motion boundaries is to segment them into regions, and then use neighboring blocks to infer motion for each region [25]. In order to avoid having to communicate the segmentation mask, [25] makes the assumption that boundaries do not change between frames, and predicts the segmentation mask from previously decoded frames.

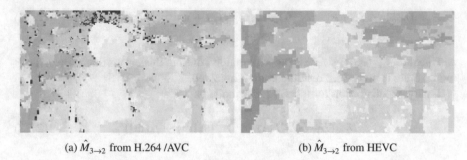

(a) $\hat{M}_{3\to 2}$ from H.264 /AVC   (b) $\hat{M}_{3\to 2}$ from HEVC

Fig. 2.9 Motion Fields (colour-coded) estimated using H.264/AVC and HEVC $INTER$ prediction, visualized using the colour-code explained in Fig. A.1. While both motion fields are reasonably smooth within objects, H.264/AVC has much more erroneous motion vectors around moving objects

The second main problem with blocks, namely their inability to represent motion that is not translational, has been addressed in various ways. Higher order motion models have been shown to be beneficial in scenes with background motion that is difficult to describe with blocks (e.g., rotation, zoom) [26, 27]. Servais et al. [28] use a content-adaptive mesh to partition the target frame into a collection of disconnected triangles, such that no triangle straddles an object boundary; this addresses the first problem of block motion mentioned above. They estimate affine motion parameters for each triangle, and hence can account for non-translational motion. While this approach is able to reduce the prediction residual, it is unclear how the motion information could be efficiently coded.

2.3   Scalable Video Compression

The goal of scalable video coding is to encode the video in an *embedded* way such that lower qualities (e.g., spatial, temporal, SNR, ...) are embedded within higher qualities; this means that at the decoder, partial streams at lower qualities can be decoded. Scalable video can be beneficial in a variety of applications.

The classic scenario of scalable video is streaming video to a set of heterogeneous clients. Because of the rapidly growing demand for consuming multimedia over networks with varying bandwidths, as well as the heterogeneity of end user devices (from smartphones to high definition displays), acceptable decoding quality can only be met if the server can instantaneously adapt the bit-rate. Figure 2.10 shows how scalable video coding can be useful if video content is streamed over heterogeneous networks.

Another scenario, which is less considered and mostly ignored in standardised video codecs, is the one of *interactive browsing and navigation* of video content, where the quality of the video progressively improves as parts of a video are revisited. Such progressive refinement is not possible if a streaming server contains multiple (unrelated) single-layer coded copies.

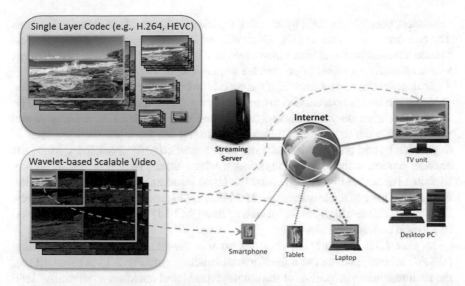

Fig. 2.10 Comparison between single layer coding and scalable video coding, in a heterogeneous video streaming application. To serve the requirements of various devices, a streaming server needs to host many copies of the same video at different qualities and resolutions. In a highly scalable video codec, on the other hand, the video is encoded in a way that lower resolutions and quality levels are *embedded* within higher qualities, such that the various requirements can be met with just a single copy of the video

Fig. 2.11 Panoramic video as an example of interactive browsing and navigation of video. The original video on the left is recorded at very high resolution, possibly recorded by a number of cameras. On a low-resolution version of the original video, the user can then select which part of the scene he or she wants to watch, which will subsequently be streamed at higher resolution

Closely related to scalability is the concept of *accessibility*, whereby the user can select an ROI in the video that will be decoded at a higher quality than the rest of the video. Accessibility requires that the codestream is organized in such a way that the amount of data that needs to be decoded outside of the ROI is minimized. A specific application where ROI coding becomes particularly appealing is the one of "panoramic" video, as shown in Fig. 2.11. An event is recorded at very high resolution, possibly obtained by stitching together a number of synchronized videos. The user can then choose an ROI, which will subsequently be streamed at higher resolution and quality. Panoramic video could also be very useful in video surveillance, where currently a number of screens are required to survey large areas.

Scalability can be best realized if the encoder works independently of the decoder. Otherwise, if the decoder decides to truncate the codestream in an unexpected manner, its state becomes desynchronized from the encoder. This leads to a phenomenon called "drift", which results in visually disturbing artefacts. As we have seen in Sect. 2.2.1, hybrid compression schemes have a predictive feedback loop, where the encoder replicates the state of the decoder, which makes such schemes inherently ill-suited for scalability. Nonetheless, the latest standardised video codecs (H.264 and HEVC) have scalable extensions, H.264/SVC [29] and SHVC [30], respectively. Section 2.3.1 gives a high-level overview of how (limited) scalability can be achieved in a hybrid coding scenario.

In Sect. 2.3.2, we turn our attention to *wavelet-based scalable video coding (WSVC)* schemes, which use a feedforward structure (see Fig. 2.1). As we will see, the multiresolution properties of the employed subband transforms "naturally" lend themselves to highly scalable video compression schemes.

2.3.1 Scalability in Hybrid Coding Schemes

Both of the latest standardised video codecs have scalable extensions, which are called H.264/SVC and SHVC. On top of spatial, temporal, and SNR scalability offered by H.264/SVC, SHVC also supports bit depth (e.g., 8 bit to 10 bit) and colour gamut scalability (e.g., BT.709 to BT.2020)[3]; since in this thesis, we focus on spatial and temporal scalability, we will not go into more details about these. As shown in [31], the gains of SHVC compared with its predecessor H.264/SVC are comparable with the gains of their non-scalable counterparts.

In the following, we present how scalability is achieved in hybrid coding schemes, and discuss the fundamental limitations that are imposed by their closed-loop nature. Temporal scalability is normally enabled by using a hierarchical B-frame structure, where the motion-compensated temporal prediction is limited to reference pictures at coarser temporal levels; an example GOP structure is shown in Fig. 2.6b. The general principle to achieve spatial and SNR (i.e., quality) scalability is to code video in multiple layers: a *base layer* (BL), which contains the lowest quality representation, plus one or more *enhancement layers* (EL), which provide improved quality of the video. Figure 2.12 shows a general structure, where the video is encoded such that two different spatial resolutions can be decoded.

The base layer can be seen as an approximation of the higher layer spatial resolutions, and can be used as prediction source in the decoding of the enhancement layers.

---

[3]BT.709 is the standard used in high definition TV (HDTV), and BT.2020 is used in ultra high definition TV (UHDTV).

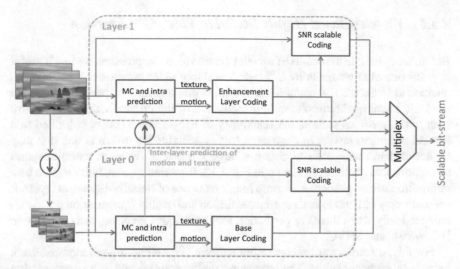

Fig. 2.12 Achieving (limited) scalability in hybrid coding schemes using enhancement layers. The video is encoded in multiple layers: a base layer (Level 0), and one or more enhancement layers (Layer 1), which improve the quality of the video. In the example, the base layer is encoded at two different spatial resolutions; the lower resolution serves as prediction reference for the higher resolution

In order to avoid the drift problem, only information from lower-numbered layers can be used to predict any given layer. While the base layer will perform identically to a single-layer coder at the same rate, the same is not true for the enhancement layers. In fact, because (lower-resolution) base-layer information is used to predict the enhancement layer, the R-D performance will be worse than the one of a single-layer coder.

One could then be tempted to include the information from all enhancement layers in the prediction of the other layers. In this case, the performance can be expected to approach the performance of a single-layer coder if all enhancement layers are consumed. However, the performance of the base layer and all intermediate layers would significantly suffer, since different estimates would be used at the encoder and the decoder, which causes the drift effect.

The discussion above highlights the fact that the feedback-loop structure of hybrid codecs is ill-suited for scalability; either the base-layer or the enhancement layer performance has to be penalized. For this reason, the number of layers is usually limited to very few. Furthermore, the quality levels have to be decided upon beforehand, and it quickly gets complicated if scalability along various dimensions (spatial, and rate) needs to be achieved.

In the next section, we turn our attention to wavelet-based scalable video compression, which is much better suited for "highly" scalable video coding since it uses a feedforward structure that is devoid of any feedback loops.

2.3.2 Wavelet-Based Highly Scalable Video Compression

Before we delve into the details of wavelet-based video compression, we find it useful to point out what we mean by a "highly" scalable video compression scheme. As mentioned in Sect. 2.1, a *scalable* bit-stream is one that may be partially discarded to obtain lower quality/resolution representation of the original video. As we have seen in the previous section, the scalability of hybrid video coders is limited to a few enhancement layers. In contrast, a highly scalable bit-stream is one that may be decoded in many ways to obtain a large number of different spatio-temporal resolutions and qualities, as shown in Fig. 2.13. Importantly, scalability should be a multi-dimensional phenomenon, not a linear sequence of causally dependent layers. It becomes very difficult to scale spatial resolution and quality (quantization precision) independently when there are prediction feedback loops involved, as is the case in H.264/SVC and SHVC.

For this, a feedforward compression scheme (i.e., without prediction feedback loop) is much better suited. The interesting multiresolution and energy compaction properties of 2D subband structures on images (see Sect. 2.1.1) have motivated research in extending them to spatio-temporal volumes, i.e., to video sequences. Karlsson and Vetterli [32] were the first to apply a 3D-DWT to videos. Because they do not use motion compensation between the frames, the temporal wavelet is only effective in regions that are *not* in motion. Furthermore, this method suffers from so-called "ghosting artefacts" in the low-pass temporal subbands.

The energy in the high-frequency subbands along the temporal domain can be significantly reduced if the frames are temporally aligned. Taubman and Zakhor [33] apply an invertible warping to frames in order to align spatial features before applying

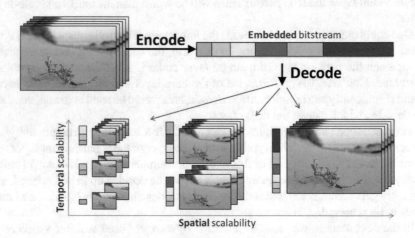

Fig. 2.13 Spatio-temporal embedding of video for highly scalable video compression. The input video is encoded in an *embedded* way, such that the video can be decoded at various frame-rates and/or spatial resolutions

a separable 3D-DWT. The proposed method is best suited for global motion, such as a panning camera, since invertible warpings cannot represent local motion of objects. Ohm [34] uses a *block displacement* scheme, where each frame is partitioned into a disjoint set of rectangular blocks, each of which can undergo translation following whole pixel motion. The 3D-DWT is then applied along the motion trajectory of the blocks. For example, the temporal lowpass subband is obtained by applying a lowpass filter along the motion trajectory of the block, followed by subsampling. The problem with this scheme is that motion trajectories of blocks might overlap (e.g., because of occlusion and contraction/expansion of objects). This means that some pixels are used multiple times in the temporal filtering, whereas others are not used at all. These "disconnected" pixels need to be treated separately in order for the whole transform to remain invertible, which negatively affects compression performance. By modifying the placement of covered and uncovered pixels depending on the motion compensation direction, Choi and Woods [35] avoid the predictive coding of covered pixels. The main problem with their approach is that for perfect reconstruction, the motion vectors have to be integer-valued. Hsiang and Woods [36] achieve half-pixel accuracy, which results in improved R-D performance compared to Choi and Woods [35].

2.3.2.1 Lifting-Based Motion Compensation

The lifting structure proposed by Sweldens [8] (see Sect. 2.1.1) was adapted to the temporal domain by several independent research groups [37–39], which led to motion compensated temporal lifting; this structure enables the construction of invertible, motion adaptive temporal transforms from arbitrary motion fields. In the literature, this structure is most often referred to as *motion-compensated temporal filtering (MCTF)* [40]. Because the DWT is not shift-invariant, the order of application of the spatial and the temporal DWT changes the final subbands. In the following, we summarize the different WSVC architectures that have been proposed in the literature.

Spatial Domain Motion Compensated Temporal Filtering "t+2D" The first, probably most natural architecture, is the "t+2D" approach, where the wavelet transform is first applied in the temporal domain, followed by a 2D spatial subband transform to the motion-compensated frames [37–39]. Let $\mathcal{W}_{i \to j}(f_i)$ denote the motion-compensated mapping of frame f_i to frame f_j. Then, using l_j and h_j to denote the low- and highpass temporal subband of the original video frames f_j, the motion-compensated lifting steps for the 5/3 biorthogonal wavelet analysis are:

$$h_k = f_{2k+1} - \frac{1}{2}\Big(\mathcal{W}_{2k \to 2k+1}(f_{2k}) + \mathcal{W}_{2k+2 \to 2k+1}(f_{2k+2})\Big)$$

$$l_k = f_{2k} + \frac{1}{4}\Big(\mathcal{W}_{2k-1 \to 2k}(h_{k-1}) + \mathcal{W}_{2k+1 \to 2k}(h_{k+1})\Big). \tag{2.23}$$

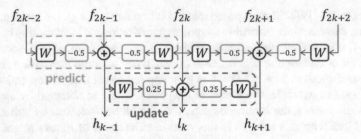

Fig. 2.14 Temporal lifting steps of a 5/3 discrete wavelet transform with motion compensation. W denotes the warping operator that maps texture from one frame to another

For clarity, Fig. 2.14 visualizes the temporal lifting steps of a 5/3 biorthogonal wavelet. The temporal wavelet transform is applied along the motion trajectories, given that the motion is correctly modelled, which significantly reduces the prediction residuals. Secker and Taubman [41] present a *lifting-based invertible motion adaptive transform (LIMAT)*, which employs a deformable mesh model for the motion that is able to model expansion and contraction of moving objects. One of the largest benefits of this scheme is that the temporal transform remains invertible, even for uninvertible motion warpings. LIMAT can benefit from bidirectional prediction provided by the 5/3 biorthogonal wavelet, whereas for the block-displacement scheme with whole pixel motion proposed in [34], the unidirectional prediction provided by the Haar wavelet performs as well. Golwelkar and Woods [42] confirm the benefits of the 5/3 wavelet. Since they work with block-based motion which has poor scalability attributes, their experimental results show that for low bit-rates, the Haar filter potentially has better R-D performance; this is due to the fact that at lower bit-rates, the block size is increased and hence larger regions of disconnected pixels are produced.

While the "t+2D" architecture has the best compression efficiency and temporal scalability, problems arise with spatial scalability. The fact that higher frequency spatial subbands are discarded at lower spatial resolutions leads to spatial aliasing. Furthermore, the motion-compensated temporal synthesis has only access to a reduced resolution version of the motion field. Rusert and Ohm [43] propose an *overcomplete* MCTF, which is similar to the "t+2D" approach, but MCTF is performed at each spatial level independently. This way, the reconstruction quality at lower spatial levels can be optimized without significantly affecting the reconstruction quality at higher spatial levels.

In-band Motion Compensation "2D+t" To address the spatial scalability problems of "t+2D" structures, Andreopoulus et al. [44] propose "in-band motion compensation". This is achieved by first applying the DWT in the spatial domain, followed by temporal filtering; because of the order the spatial and temporal DWT are applied, this structure is often referred to as "2D+t". With this structure, separate motion fields can be employed for different subbands, which mitigates the problem of poor scalability inherent to block-based motion models. The main weakness of this approach is that it tends to produce artefacts at lower temporal resolutions. It also has worse compression efficiency than the "t+2D" architecture because motion information for each spatial level is estimated and coded independently, without exploiting correlations between the motion vectors of different subbands.

Adaptive Schemes The inherent problems of the two previous architectures have been addressed by adaptive schemes [45, 46]. Mehrseresht and Taubman [45] propose an adaptive spatio-temporal decomposition that continuously adapts between the "t+2D" and the "2D+t" structure. This can remove artefacts due to motion failure for both reduced spatial and temporal resolutions. Their adaptive scheme solves the problem of spatial aliasing artefacts due to misalignments, while almost preserving the compression efficiency of the "t+2D" structure. This is achieved by estimating the local performance of the motion model, which is assumed to be proportional to the energy in the high-pass temporal frames. The experimental results, although performed only on two video sequences, indicate that the transform can significantly improve the compression efficiency needed for spatial and temporal scalability. Similar to [45], Gao et al. [46] use the fact that there is more energy in highpass subbands in places where the motion model fails. They note that if mismatches occur, they do so simultaneously in all highpass subbands (i.e., HL, LH, HH), and use this fact to reduce computational complexity in the prediction process. The downside of this approach is that MCTF prediction needs to be performed twice in order to detect motion mismatches.

Even though there have been significant improvements in the field of scalable video coding, the rate-distortion performance of the above-mentioned WSVC schemes remains inferior to H.264/SVC (and SHVC for that matter). One reason for this is the fact that WSVC has problems at *discontinuities* in the motion field [47]; band-limited sampling of motion fields smooths out the sharp transitions at moving object boundaries, which creates non-physical motion. Lalgudi et al. [48] adopt a similar approach to the LIMAT framework [41] to compress volume rendered images. They determine the underlying geometric relationship between volume rendered images, which is then incorporated into the lifting steps of a temporal wavelet transform. Experimental results show better compression performance than H.264/AVC at much lower computational complexity than normal motion compensation. Their results suggest that MCTF can produce excellent results if motion discontinuities are properly handled. Garbas et al. [49] show similarly promising results in the context of wavelet-based multiview coding. They apply a 4D-DWT (3D spatiotemporal, plus 1D for disparities), and observe that the temporal correlation characteristics between neighbouring views are almost identical. Similarly, over a small time instance, the view correlation is nearly constant. In both methods, the motion discontinuities are properly handled because of some intrinsic properties of the setup, which motivates the incorporation of motion discontinuities into the spatio-temporal wavelet transform.

2.3.3 Scalable Coding of Motion and Motion Discontinuities

One of the main contributions lies in the proposal of effective ways of handling problematic regions around moving objects. For this, the methods we propose employ with piecewise-smooth motion fields with discontinuities around moving object boundaries. Unlike block motion, such "physical" motion scales naturally, except at the discontinuous motion boundaries. In this section, we show how motion fields with discontinuities at moving object boundaries can be coded efficiently and scalably. The estimation of such motion fields is a challenging field of active research on its own; Sect. 3.1.2 reviews motion estimation schemes that aim at estimating appropriate motion fields. Quite apart from the difficulties associated with the estimation of "true" motion of objects in a scene, a significant obstacle to the use of such "optical flow" fields for video coding is their (potentially) high communication cost.

An early attempt to use optical flow in a video coder was made by Krishnamurthy et al. [50], who propose to use a multiscale optical flow based motion estimator [51], which estimates a *smooth* dense flow; each pixel in the *target* frame gets (potentially) assigned a different motion vector. The authors show that the smooth motion fields estimated by their method are compressible, as long as the motion is not too complicated. More recently, dense motion estimation methods for video coding have been proposed which optimise for both smoothness and compressibility of the motion field [52]. As the authors point out, modern optical flow algorithms favour sharp discontinuities at moving object boundaries, which makes them harder to be compressed. Zheng et al. [53] estimate a piecewise-smooth motion field using hierarchical block matching; the motion field is coded using a modification of the depth intra coding algorithm which is part of 3D-HEVC [54]. They show comparable compression performance to HEVC, while overcoming the block artefacts inherent to any block motion based video compression system. Young et al. [55] *explicitly* handle motion discontinuities, and advocate "compression-regularized optical flow", where piecewise-smooth motion fields are estimated, and discontinuities are explicitly coded. As shown in [56], such motion fields are highly scalable. In the following section, we present how the wavelet bases can be adapted in the vicinity of (motion) discontinuities, in order to make motion fields more compressible.

2.3.3.1 Scalable Representation of Discontinuities Using Breakpoints

Mathew and Taubman [57] propose a compelling way of handling discontinuities. In their case, they work on depth maps, and discontinuities appear at object boundaries. They propose a scheme that incorporates a scalable and embedded representation of object boundaries using *breakpoints*, which are encoded in a separate pyramid structure. The presence of breakpoints is determined in an R-D optimization framework. They propose a *breakpoint-adaptive DWT (BPA-DWT)*, which uses breakpoints to avoid wavelet bases from crossing discontinuity boundaries. This results in a reduction of the magnitude of subband samples in the vicinity of discontinuities, which mitigates the spatial scalability artefacts. Experimental results show an improvement over JPEG2000 for encoding depth maps; in particular, ringing artefacts around discontinuities are significantly reduced. In this thesis, we make extensive use of breakpoints to efficiently code motion fields; that is, the breakpoints are used to drive a BPA-DWT on the horizontal and vertical component of the motion fields.

The technical details on how breakpoints are estimated in an R-D optimized way can be found in [57]. In the next section, we summarise how geometry information can be induced from an existing breakpoint field, which is relevant for this thesis.

2.3.3.2 Spatial Induction of Breakpoints

Breakpoints are organized in a hierarchical manner, such that breakpoints at finer spatial levels can be *induced* from coarser levels.

The breakpoint field at spatial level η consists of squares of size $2^\eta \times 2^\eta$ called *cells*, which are the fundamental unit used for inducing discontinuities. A cell consists of four *perimeter arcs* (cyan lines in Fig. 2.15), as well as two *root arcs* (grey lines in Fig. 2.15). The significance of root arcs is that they do not exist at coarser levels in the pyramid. Each arc can contain at most one breakpoint. If a cell contains exactly two perimeter breakpoints, and the root arcs at this level have no explicitly coded breaks, connecting the two perimeter breaks *induces* breakpoints onto the root arcs. To avoid confusion, we use the term *vertices* to identify the explicitly coded breaks. What this means then is that spatial induction transfers discontinuity information recursively from coarser level vertices to finer levels in the hierarchy, except where such transfer would be in conflict with finer level vertices. Since each arc in the hierarchy may have a coded vertex, the breakpoint representation is described by a vertex field that is scalable in precision. At lower bit-rates, the representation is necessarily highly sparse, with most breaks being induced.

Even so, however, signalling a sparse set of vertices at low precision can still occupy a significant portion of the bit-rate budget in some video compression

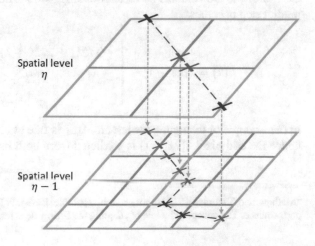

Fig. 2.15 Highly scalable geometry representation: Two breakpoints on the *perimeter* of the same *cell* (cyan squares) can induce discontinuity information onto the *root* arcs (grey lines); such spatially induced breaks are indicated by purple crosses. If the root arc contains a vertex (red cross), the inducing is stopped

Spatial level η

Spatial level $\eta - 1$

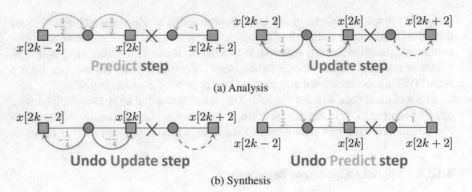

(a) Analysis

(b) Synthesis

Fig. 2.16 Lifting implementation of a 1D breakpoint-adaptive DWT. Breakpoints are used to modify the lifting steps around discontinuities

applications. In this thesis, we extend the existing spatial breakpoint induction to a spatio-temporal breakpoint induction (see Sect. 5.4), which can be used to complete/improve breakpoint fields at finer temporal scales with breakpoint information from coarser temporal levels.

2.3.3.3 Breakpoint-Adaptive DWT

In this section, we explain in the 1D case how breakpoints can be used to modify the lifting steps of the 5/3 biorthogonal DWT to efficiently code piecewise-smooth motion fields; the extension to 2D is straightforward by separable extension.[4] We use Fig. 2.16 to guide the description.

Let A_k denote the *arc* comprising the pixels $x[2k]$, $x[2k + 1]$, and $x[2k + 2]$. Furthermore, we use $A_k = -1$ to indicate that the arc contains a breakpoint between pixels $2k$ and $2k + 1$; similarly, $A_k = +1$ if there is a breakpoint between $2k + 1$ and $2k + 2$. If the arc contains no breakpoints, $A_k = 0$. Then, the breakpoint-adaptive predict step becomes

$$h[k] = x[2k + 1] - \begin{cases} \frac{1}{2}(x[2k] + x[2k + 2]) & A_k = 0 \\ x[2k + 2] & A_k = -1 \\ x[2k] & A_k = 1 \end{cases} \quad (2.24)$$

In the example of the figure, the left arc A_{k-1} is free of breakpoints ($A_{k-1} = 0$), and hence the odd pixel $x[2k - 1]$ is predicted from both its parents. The right arc, on

---

[4]Mathew and Taubman [57] propose a non-separable BPA-DWT, which has slightly improved performance. The general principle of adapting the lifting steps remains the same.

the other hand, contains a breakpoint between $x[2k]$ and $x[2k+1]$ ($A_k = -1$), and therefore the odd pixel $x[2k+1]$ is predicted only from its "right" parent $x[2k+2]$.

As proposed in [57], the *update* step is *disabled* if there is a breakpoint present on the arc. That is,

$$l[k] = x[2k] + \frac{1}{4}\left(\mathcal{I}(A_{k-1})x[2k-1] + \mathcal{I}(A_k)x[2k+1]\right), \qquad (2.25)$$

where $\mathcal{I}(A_k) = 1$ if $A_k = 0$, and zero otherwise. In the figure, the pixel at location $x[2k]$ is therefore only updated from $x[2k-1]$, since A_k contains a breakpoint (e.g., $\mathcal{I}(A_k) = 0$), which disables the update step on that arc. The synthesis simply undoes the update and predict step, again using the modified weights, as shown in Fig. 2.16b.

The effect of the BPA-DWT can be appreciated in Fig. 2.17, where we show a 1-level spatial DWT, and compare it with the corresponding breakpoint-adaptive

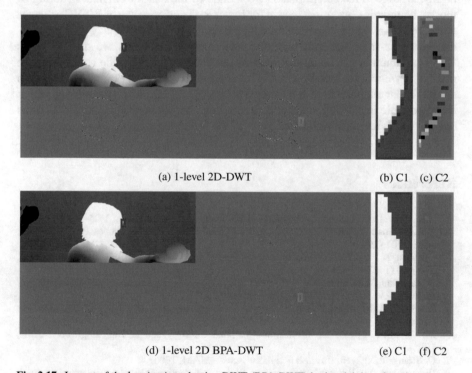

(a) 1-level 2D-DWT   (b) C1   (c) C2

(d) 1-level 2D BPA-DWT   (e) C1   (f) C2

Fig. 2.17 Impact of the breakpoint-adaptive DWT (BPA-DWT) in the vicinity of motion discontinuities. **a** Shows a 1-level 2D-DWT *without* breakpoints, where large wavelet coefficients can be observed around motion discontinuities, as evidenced in the crops (**b**) and (**c**); **d** shows the 1-level 2D breakpoint-adaptive DWT; the corresponding crops in **e** and **f** show how the magnitude of the coefficients around discontinuities in the motion field is greatly reduced by the use of breakpoints. For visualization purposes, the values of the coefficients have been clipped at ± 5

implementation. One can see how the magnitude of the wavelet coefficients is significantly reduced around motion discontinuities.

2.4   Summary

In this first of two background chapters, we introduced the fundamentals of video compression with a focus on scalability, which is of particular interest to this thesis. The predictive feedback loop present in hybrid coding schemes makes them inherently ill-suited for scalability. While the need for scalability has been acknowledged by the introduction of scalable extensions in the two latest standardised codecs, their scalability is limited to only a few number of layers. The feedforward structure of wavelet-based scalable video compression schemes makes them better suited for highly scalable video compression. We identified the lack of efficient handling of motion boundaries as a prime reason for lower R-D performance of WSVC schemes compared than single-layer coding, and presented an attractive way of compressing piecewise-smooth motion with discontinuities. In this thesis, we show how this "physical" motion can be used to perform better motion inference across time. In particular, we will see how it naturally lends itself to perform TFI if all residual data at a particular temporal level is quantized to zero. This "intrinsic" upsampling, which is not possible in existing video codecs, is a key contribution of this thesis. For this reason, we review existing TFI methods in the following chapter.

References

1. R.M. Gray, D.L. Neuhoff, Quantization. IEEE Trans. Inf. Theory **44**(6), 2325–2383 (1998)
2. D.A. Huffman, A method for the construction of minimum- redundancy codes. Proc. IRE **40**(9), 1098–1101 (1952)
3. I. Witten, R. Neal, J. Cleary, Arithmetic coding for data compression. Commun. ACM **30**(6), 520–540 (1987)
4. D. Taubman, M.W. Marcellin, *JPEG2000: Image Compression Fundamentals, Standards, and Practice* (Kluwer Academic Publishers, Boston, 2002)
5. N. Ahmed, T. Natarajan, K.R. Rao, Discrete cosine transform. IEEE Trans. Comput. **23**, 90–93 (1974)
6. M. Vetterli, J. Kovačević, *Wavelets and Subband Coding* (Prentice Hall, Englewood Cliffs, 1995)
7. G. Strang, T. Nguyen, *Wavelets and Filter Banks* (SIAM, 1996)
8. W. Sweldens, The lifting scheme: a custom-design construction of biorthogonal wavelets. Appl. Comput. Harmon. Anal. **3**(2), 186–200 (1996)
9. W. Sweldens, The lifting scheme: a construction of second generation wavelets. SIAM J. Math. Anal. **29**(2), 511–546 (1998)
10. D. Le Gall, A. Tabatabai, Sub-band coding of digital images using symmetric short Kernel filters and arithmetic coding techniques, in Proceedings of the IEEE International Conference on Acoustics, Speech, and Signal Processing (1988), pp. 761–764
11. J.M. Shapiro, Embedded image coding using zerotrees of wavelet coefficients. IEEE Trans. Image Proc. **41**(12), 3445–3462 (1993)

12. A. Said, W.A. Pearlman, A new, fast, and efficient image codec based on set partitioning in hierarchical trees. IEEE Trans. Circ. Syst. Video Tech. **6**(3), 243–250 (1996)
13. D. Taubman, High performance scalable image compression with EBCOT. IEEE Trans. Image Proc. **9**(7), 1158–70 (2000)
14. T. Wiegand, G.J. Sullivan, G. Bjøntegaard, A. Luthra, Overview of the H.264/AVC video coding standard. IEEE Trans. Circ. Syst. Video Tech. **13**(7), 560–576 (2003)
15. G.J. Sullivan, J.-R. Ohm, W.-J. Han, T. Wiegand, Overview of the high efficiency video coding (HEVC) standard. IEEE Trans. Circ. Syst. Video Tech. **22**(12), 1649–1668 (2012)
16. N.S. Jayant, P. Noll, *Digital Coding of Waveforms: Principles and Applications to Speech and Video*. Prentice Hall Professional Technical Reference (1990), ISBN: 0132119137
17. H. Schwarz, D. Marpe, T. Wiegand, Analysis of Hierarchical B-Pictures and MCTF (2006), pp. 1929–1932
18. J.R. Jain, A.K. Jain, Displacement measurement and its application in interframe image coding. IEEE Trans. Comm. **29**(12), 1799–1808 (1981)
19. G.J. Sullivan, J.-R. Ohm, W.-J. Han, T. Wiegand, Overview of the high efficiency video coding. IEEE Trans. Circ. Syst. Video Tech. **22**(12), 1649–1668 (2012)
20. G. Bjøntegaard, Calculation of average PSNR Differences between RD-curve, VCEG-M33, ITU-T SG16/Q6, Technical Report (2001)
21. T. Wiegand, H. Schwarz, A. Joch, F. Kossentini, G.J. Sullivan, Rate-constrained coder control and comparison of video coding standards. IEEE Trans. Circ. Syst. Video Tech. **13**(7), 688–703 (2003)
22. G.J. Sullivan, T. Wiegand, Rate-distortion optimization for video compression. IEEE Sig. Proc. Mag. **15**(6), 74–90 (1998)
23. M. Chan, Y. Yu, A. Constantinides, Variable size block matching motion compensation with applications to video coding **137**(4), 205–212 (1990)
24. R. Mathew, D. Taubman, Quad-tree motion modeling with leaf merging. IEEE Trans. Circ. Syst. Video Tech. **20**(10), 1331–1345 (2010)
25. M.T. Orchard, Predictive motion-field segmentation for image sequence coding. IEEE Trans. Circ. Syst. Video Tech. **3**(1), 54–70 (1993)
26. A. Glantz, A. Krutz, T. Sikora, Adaptive global motion temporal prediction for video coding, in *Picture Coding Symposium* (2010)
27. M. Tok, V. Eiselein, T. Sikora, Motion modeling for motion vector coding in HEVC, in *Picture Coding Symposium* (2015)
28. M. Servais, T. Vlachos, T. Davies, Affine motion compensation using a content-based mesh **152**(4), 415–423 (2005)
29. H. Schwarz, D. Marpe, T. Wiegand, Overview of the scalable video coding extension of the H.264/AVC standard. IEEE Trans. Circ. Syst. Video Tech. **17**(9), 1103–1120 (2007)
30. P. Helle, H. Lakshman, M. Siekmann, J. Stegemann, T. Hinz, H. Schwarz, D. Marpe, T. Wiegand, A scalable video coding extension of HEVC, in *Proceedings of the IEEE Data Compression Conference* (2013)
31. J. Boyce, Y. Ye, J. Chen, A. Ramasubramonian, Overview of SHVC: scalable extensions of the high efficiency video coding (HEVC) standard, in *IEEE Transactions on Circuits and Systems for Video Technology* (2015)
32. G. Karlson, M. Vetterli, Three-dimensional subband coding of video, in *Proceedings of IEEE International Conference on Acoustics, Speech, and Signal Processing* (1988)
33. D. Taubman, A. Zakhor, Multirate 3-D subband coding of video. IEEE Trans. Image Proc. **3**(5), 572–88 (1994)
34. J.-R. Ohm, Three-dimensional subband coding with motion compensation. IEEE Trans. Image Proc. **3**(5), 559–571 (1994)
35. S.J. Choi, J.W. Woods, Motion-compensated 3-D subband coding of video. IEEE Trans. Image Proc. **8**(?), 155–67 (1999)
36. S.T. Hsiang, J.W. Woods, Invertible three-dimensional analysis/ synthesis system for video coding with half-pixel-accurate motion compensation, in *SPIE Conf. on Vis. Commu. and Im. Proc.* (1999), pp. 537–546

37. A. Secker, D. Taubman, Motion-compensated highly scalable video compression using an adaptive 3D wavelet transform based on lifting, in *Proceedings of IEEE International Conference on Image Processing* (2001), pp. 1029–1032
38. B. Pesquet-Popescu, V. Bottreau, Three-dimensional lifting schemes for motion compensated video compression, in *Proceedings of IEEE International Conference on Acoustics, Speech and Signal Processing* (2001), pp. 1793–1796
39. L. Luo, J. Li, S. Li, Z. Zhuang, Y.-Q. Zhang, Motion compensated lifting wavelet and its application in video coding, in *Proceedings of IEEE International Conference on Multimedia and Expo* (2001)
40. J.-R. Ohm, M. Schaar, J. Woods, Interframe wavelet coding - motion picture representation for universal scalability. Sig. Proc.: Im. Comm. **19**(9), 877–908 (2004)
41. A. Secker, D. Taubman, Highly scalable video compression with scalable motion coding. Proc. IEEE Int. Conf. Image Proc. **13** (2004)
42. A. Golwelkar, J.W. Woods, Scalable video compression using longer motion compensated temporal filters. Vis. Comm. Im. Proc. **5150**, 1406–1416 (2003)
43. T. Rusert, J.-R. Ohm, Overcomplete MCTF for improved spatial scalability in 3D wavelet video compression. SPIE Conf. Vis. Comm. Im. Proc. **5960** (2005)
44. Y. Andreopoulos, A. Munteanu, J. Barbarien, M. Van der Schaar, J. Cornelis, P. Schelkens, In-band motion compensated temporal filtering. Sig. Proc: Im. Comm. **19**(7), 653–673 (2004)
45. N. Mehrseresht, D. Taubman, An efficient content-adaptive motion-compensated 3-D DWT with enhanced spatial and temporal scalability. IEEE Trans. Image Proc. **15**(6), 1397–412 (2006)
46. A. Gao, N. Canagarajah, D. Bull, Adaptive in-band motion compensated temporal filtering based on motion mismatch detection in the highpass subbands. Int. Symp. Visual Comm. Image Proc. **6077** (2006)
47. N. Adami, A. Signoroni, R. Leonardi, State-of-the-art and trends in scalable video compression with wavelet-based approaches. IEEE Trans. Circ. Syst. Video Tech. **17**(9), 1238–1255 (2007)
48. H.G. Lalgudi, M.W. Marcellin, A. Bilgin, H. Oh, M.S. Nadar, View compensated compression of volume rendered images for remote visualization. IEEE Trans. Image Proc. **18**(7), 1501–11 (2009)
49. J.-U. Garbas, B. Pesquet-Popescu, A. Kaup, Methods and tools for wavelet-based scalable multiview video coding. IEEE Trans. Circ. Syst. Video Tech. **21**(2), 113–126 (2011)
50. R. Krishnamurthy, J.W. Woods, P. Moulin, Frame interpolation and bidirectional prediction of video using compactly encoded optical-flow fields and label fields. IEEE Trans. Circ. Syst. Video Tech. **9**(5), 713–726 (1999)
51. P. Moulin, R. Krishnamurthy, J.W. Woods, Multiscale modeling and estimation of motion fields for video coding. IEEE Trans. Image Proc. **6**(12), 1606–1620 (1997)
52. G. Ottaviano, P. Kohli, Compressible motion fields, in *Proceedings IEEE Conference on Computer Vision and Pattern Recognition* (2013)
53. A. Zheng, Y. Yuan, H. Zhang, H. Yang, P. Wan, O. Au, Motion vector fields based video coding, in *Proceedings IEEE International Conference on Image Processing* (2015), pp. 2095–2099
54. K. Müller, H. Schwarz, D. Marpe, C. Bartnik, S. Bosse, H. Brust, T. Hinz, H. Lakshman, P. Merkle, F.H. Rhee et al., 3D high-efficiency video coding for multi-view video and depth data. IEEE Trans. Image Proc. **22**(9), 3366–3378 (2013)
55. S. Young, R. Mathew, D. Taubman, Joint estimation of motion and arc breakpoints for scalable compression, in *Proceedings IEEE Global Conference on Signal and Information Processing* (2013)
56. S. Young, R. Mathew, D. Taubman, Embedded coding of optical flow fields for scalable video compression, in *IEEE International Workshop on Multimedia Signal Processing* (2014)
57. R. Mathew, D. Taubman, P. Zanuttigh, Scalable coding of depth maps with R-D optimized embedding. IEEE Trans. Image Proc. **22**(5), 1982–95 (2013)

Chapter 3
Temporal Frame Interpolation (TFI)

As discussed in the last chapter, the temporal scalability of standardized scalable video codecs, e.g., H.264/SVC [1] and SHVC [2], is limited to reducing the framerate. One of the main reasons that the framerate can not be increased is that the target-frame anchored motion is estimated in an opportunistic way, which means that it does not in general describe the "true" motion trajectory of objects in the scene. In contrast, in the motion anchoring strategies explored in this thesis, motion information is anchored at reference frames, and *temporal frame interpolation (TFI)* is the essential building block that allows us to form predictions of the target frames.

TFI, or *framerate up-conversion* (FRUC), is used to increase the framerate of a video by inserting frames in between existing frames. The ability to increase the framerate of a video has a variety of practical applications. TFI is an integral part in displays, which typically are driven at a higher framerate than the recorded video, to reduce motion blur [3]; furthermore, sequential colour displays – notably *liquid crystal on silicon* (LCOS) and micro-mirror projectors – display only one colour plane at a time, which leads to motion-colour artefacts without high quality TFI. In distributed video coding [4], temporally interpolated frames are used as side-information for Wyner-Zyv decoding. In video post-processing, TFI is used to create slow-motion effects.

Current state-of-the-art TFI methods consist of two main steps: First, the motion between two neighbouring frames is estimated; Second, the estimated motion is used to interpolate the temporally upsampled frame. We describe common motion estimation schemes used in TFI in Sect. 3.1, and then review state-of-the-art TFI schemes in Sect. 3.2.

© Springer Nature Singapore Pte Ltd. 2018
D. Rüfenacht, *Novel Motion Anchoring Strategies for Wavelet-Based Highly Scalable Video Compression*, Springer Theses,
https://doi.org/10.1007/978-981-10-8225-2_3

3.1   True Motion Estimation

As we have seen in Sect. 2.2.1, the motion estimation in a video compression system does not attempt to estimate motion vectors that follow the "true" motion trajectory; instead, the "motion" is chosen so as to minimize the total number of bits required to code the motion *and* prediction residual of the frame textures. In order to distinguish motion estimation schemes that are used in video compression from the ones used in video processing applications such as TFI, the latter are referred to as *true motion estimation (TME)* algorithms.

In this section, we give an overview of the TME methods that are popular in TFI schemes. While much less used due to their relatively high computational complexity compared to simple block matching schemes, we also give an overview of optical flow methods. We focus on motion-discontinuity preserving optical flow methods, which is what we employ in the methods presented in this thesis.

3.1.1   Block Matching Algorithms

With the goal of improving block motion fields for TFI, a variety of algorithms have been proposed that aim at estimating smoother motion fields using block motion. Smoothness can be achieved both *implicitly* and *explicitly*; often, TME schemes combine both implicit and explicit ways. A common way of implicitly imposing smoothness is achieved by multiresolution block matching, where the coarser levels impose smoothness on the finer levels. Unlike explicit ways, the added advantage of such hierarchical approaches is that they can help speeding up the motion estimation search.

A widely used TME method is the so-called *3D recursive search (3DRS)* proposed by Haan et al. [5]. 3DRS uses spatio-temporally neighbouring blocks as motion vector candidates in order to accelerate the convergence of the algorithm and to implicitly impose smoothness in the motion field. They further apply a median filter, which can be seen as an explicit way of imposing smoothness. While it is able to produce smoother motion fields, problems arise at moving object boundaries with large motion differences. Beric et al. [6] show that this problem can be mitigated by applying multiple passes of 3DRS.

Ha et al. [7] propose *overlapped block motion estimation (OBME)*, where, as the name suggests, the idea is to overlap blocks in the block matching process. This can be achieved by increasing the block size of the matching function in (2.14), while keeping the actual blocks the same size. Since this inevitably increases the computational complexity, they propose to perform the motion search on a sub-sampled grid.

Another way of achieving smoothness explicitly is by adding a *penalty term* (smoothness constraint) to the block distortion measure of (2.14), as follows:

$$E(\mathbf{u}) = \sum_{\mathbf{m} \in K} \rho_1 \left(f_a[\mathbf{m} + \mathbf{u}], \, f_b[\mathbf{m}] \right) + \lambda \sum_{\mathbf{v} \in N_K} \rho_2 \left(\mathbf{u}, \mathbf{v} \right), \qquad (3.1)$$

where $\mathbf{v}$ denotes the motion vectors in the neighbourhood N_K of block K (causal neighbours). We further subscript the matching criterion ρ_j to emphasize that different matching criteria can be used for the data and the penalty term. For example, Lu et al. [8] use the L1-norm, or *sum of absolute differences (SAD)*, for the data and the L2 norm for the penalty term. The use of SAD for the data term is quite common in TFI schemes because of its lower computational complexity. It is worth noting that the quality of the motion field can be greatly improved by adding a penalty term. However, it also significantly increases the computational complexity of the motion estimation stage compared to (hierarchical) block matching. In fact, as we will see in Sect. 3.1.2, the formulation becomes similar to what is employed in optical flow algorithms, while still suffering from artificial discontinuities at block boundaries.

Figure 3.1 shows motion fields estimated using block-matching *without* smoothness constraint (Fig. 3.1a), with smoothness constraint imposed by state-of-the-art video coders (Fig. 3.1b, c), as well as with explicit smoothness constraint obtained from the state-of-the-art TFI scheme by Lu et al. [8] (Fig. 3.1d). One can see how the block motion field is quite noisy without smoothness constraints, which is not

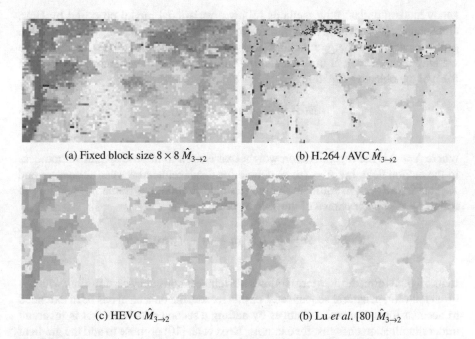

(a) Fixed block size $8 \times 8 \ \hat{M}_{3 \rightarrow 2}$ (b) H.264 / AVC $\hat{M}_{3 \rightarrow 2}$

(c) HEVC $\hat{M}_{3 \rightarrow 2}$ (b) Lu *et al.* [80] $\hat{M}_{3 \rightarrow 2}$

Fig. 3.1 Comparison of motion fields estimated using **a** unconstrained block matching using a fixed block size of 8×8, **b** H.264/AVC, **c** HEVC, and **d** with explicit smoothing constraints and occlusion handling using Lu et al.'s [8] TFI method. Motion fields are visualized using the colour-code explained in Fig. A.1

well-suited for TFI. The tree-structured block matching schemes employed in video coders inherently favour smoothness within objects. As a matter of fact, the formulations for finding motion vectors in a compression scheme (2.19), and the explicit smoothness constraint used in TMEs (3.1), are very similar. The main difference can be observed in occluded regions, where the *opportunistic* nature of motion estimated in a video coder can be observed. More precisely, in Fig. 3.1b, one can observe how the motion field estimated by H.264/AVC contains many motion vectors that point in "arbitrary" directions around moving objects. The motion field produced by HEVC (Fig. 3.1c) looks much "cleaner"; however, closer inspection shows a lot of zero motion vectors in such regions (white blocks). The explicit smoothness handling of Lu et al. [8], coupled together with reasoning about occluded regions, leads to improved motion around moving objects, as well as smoother motion within moving objects. TFI schemes that are based on block motion in general have problems in representing non-translational motion, such as rotation and zoom. In the next section, we present optical flow, which aims at estimating a pixel-wise motion field.

3.1.2 Optical Flow

Many modern optical flow methods follow a variational model proposed by Horn and Schunck [9], and pose the motion estimation problem as an energy minimization problem, where the energy function consists of a data and a smoothness term. Before we have a closer look at the data and the smoothness term, we present the general form of the objective function:

$$E(\mathbf{u}) = \underbrace{E_D(\mathbf{u})}_{\text{data}} + \lambda \cdot \underbrace{E_S(\mathbf{u})}_{\text{smoothness}} , \tag{3.2}$$

where $\lambda > 0$ is the regularization weight that is used to impose spatial smoothness in the motion field. Let $\mathbf{u} = (u, v)$ denote the displacement between frames f_a and f_b; that is, we seek to estimate the motion field anchored at frame f_a, pointing to frame f_b. A popular choice of the *data term* is:

$$E_D(\mathbf{u}) = \sum_{\mathbf{m}} \|f_b(\mathbf{m} + \mathbf{u}) - f_a(\mathbf{m})\|. \tag{3.3}$$

In its original form, the data term follows a brightness constancy assumption, which is often not valid in natural sequences. For this reason, the data term has been extended to account for illumination changes by adding a second data term that is invariant under illumination changes; for example, Brox et al. [10] propose to add the gradient of the image. That is,

$$E_D(\mathbf{u}) = \sum_{\mathbf{m}} \frac{1}{2}\|f_b(\mathbf{m} + \mathbf{u}) - f_a(\mathbf{m})\| + \frac{1}{2}\beta\|\nabla f_b(\mathbf{m} + \mathbf{u}) - \nabla f_a(\mathbf{m})\|, \tag{3.4}$$

where ∇ is the discrete gradient operator, and β is a weight that balances the two matching costs of the data term.

The *smoothness term* is typically designed to be *edge-preserving* [11, 12].

$$E_S(\mathbf{u}) = \sum_{\mathbf{m}} w(\mathbf{m})||\nabla \mathbf{u}(\mathbf{m})||, \qquad (3.5)$$

where $||\nabla \mathbf{u}(\mathbf{m})||$ is the total variation (TV) regulariser, and $w(\mathbf{m}) = exp(-||\nabla f_a||)$ is a structure adaptive map that aims at maintaining motion discontinuity.

Over the last three decades, there has been an impressive amount of work on improving the classical Horn–Schunck objective. In the interest of conciseness, we fast-forward to the most recent, best-performing optical flow methods, with a particular focus on optical flow methods that preserve motion discontinuities, since this is the type of motion fields the subsequent work in this thesis requires.

Xu et al.'s [13] *motion detail preserving (MDP)* optical flow algorithm uses an extended coarse-to-fine refinement framework, which is able to recover motion details at each scale by reducing the reliance of flow estimates that are propagated from coarser scales. Large displacements are handled by using sparse feature detection and matching, and a dense nearest-neighbour patch matching algorithm is used to handle small textureless regions which are likely missed by the feature matching algorithm. Furthermore, an adaptive structure map which maintains motion discontinuity is used in the optical flow regularization term.

Wulff and Black [14] propose a layered motion model, which is able to obtain piecewise-smooth motion fields with sharp discontinuities on sequences that are heavily affected by motion blur. While currently limited to two layers, the authors show that the scheme is quite widely applicable. Revaud et al. [15] propose an *edge-preserving interpolation of correspondences for optical (EPIC)* flow. They estimate a sparse set of correspondences between the two frames, also referred to as "features", which are then interpolated in an edge-preserving way to a dense motion field. While this approach is able to account for large displacements, the fact that only the finest spatial resolution is used means that the method is prone to errors in regions of repetitive textures.

The edge-preserving sparse-to-dense interpolation proposed by Revaud et al. [15] has been used by various other recent state-of-the-art optical flow methods to go from a set of sparse matches to a dense motion field, while preserving edges. Menze et al. [16] propose "discrete flow", where they conjecture that computing the integer part of the motion is the hardest problem, and formulate optical flow estimation as a *discrete* inference problem in a conditional random field. EPIC flow is then used to obtain sub-pixel flow and extrapolate motion in occluded regions. Chen and Koltun [17] propose "full flow", where they optimize the classical Horn–Schunck objective [9] globally over full regular grids. Interestingly, they show that large displacements can effectively be handled without any feature matching, as used for example in [15].

We end this section with a motivating example of the performance of state-of-the-art optical flow methods. Figure 3.2 shows estimated motion fields for a very

(a) Ambush 3 $\frac{1}{2}(f_{25} + f_{26})$ (b) Ground Truth $\hat{M}_{25\to26}$

(c) Horn and Schunck [81] $\hat{M}_{25\to26}$ (d) MDP-2 [85] $\hat{M}_{25\to26}$

(e) EPIC flow [87] $\hat{M}_{25\to26}$ (f) Full flow [89] $\hat{M}_{25\to26}$

Fig. 3.2 Comparison of various optical flow methods on a frame from the Sintel test dataset "Ambush 3". **a** shows the average of the two input frames; **b** shows the ground truth motion field. **c–f** show the estimated motion fields obtained using Horn and Schunck [9], Xu et al.'s motion detail preserving optical flow (MDP-2) [13], Revaud et al.'s edge preserving interpolation of correspondences (EPIC) flow [15], and Chen and Koltun's full flow [17], respectively. Motion fields are visualized using the colour-code explained in Fig. A.1. Images from http://sintel.is.tue. mpg.de/results

challenging sequence which contains large motion that is affected by motion blur and atmospheric effects. One can appreciate the significant improvements that have been achieved in recent years.

3.2 State-of-the-Art in TFI

The field of TFI has been an active field of research for over two decades. With the ever growing increase of both spatial and temporal resolution of video content, the use of good motion fields, as well as a proper handling of occluded regions, has become ever more important. In the following, we give an overview of TFI methods, with a specific focus on TFI schemes that aim at high quality interpolation; for a comprehensive overview of fast TFI schemes, the interested reader is referred to [18].

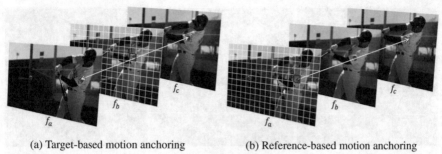

(a) Target-based motion anchoring   (b) Reference-based motion anchoring

Fig. 3.3 Comparison of target-based and reference-based anchoring of motion. **a** shows the target-based motion anchoring, where for each block in the target frame, the best matching block in the reference frames is found; **b** shows the reference-based motion anchoring, where for each block in one of the reference frames, the best matching block in the other reference frame is found. Appropriate scaling is used to map the block to the target frame

When designing a TFI method, there are a number of fundamental design choices that affect the performance of the scheme. One choice is whether a block-based or an optical flow motion estimation scheme is to be used. If block-based motion is used, many schemes use *overlapped block motion compensation (OBMC)* to reduce block artefacts; as its name suggests, each pixel of the target frame is predicted as a linear combination of the estimates given by motion vectors of its block as well as neighbouring blocks. It is worth highlighting that while OBME (see Sect. 3.1.1) creates smoother, more accurate motion vectors, OBMC creates smoother textures since texture information from multiple locations is averaged together. Next, it has to be decided whether the motion estimation is performed at the reference or at the target frame; this is visualized in Fig. 3.3. Another important decision is whether real-time performance is required or not, which directly impacts the quality of motion estimation as well as complexity of texture optimizations.

Before we start with the review, we point out an important feature of TFI schemes, which is how regions around moving objects are handled. We use Fig. 3.4 to give more insight into the problem of occlusions and disocclusions. The figure shows two reference frames f_a and f_c, and the non-existing target frame f_b (in grey). In the sequence, the moving object (MO) is a hand which lifts an apple, and the background (BG) is static. Around the MO, certain regions get "forward" disoccluded (uncovered) as we transition from f_a to f_b (yellow in Fig. 3.4). Likewise, there are regions that get "reverse" disoccluded if we transition (in reverse direction) from frame f_c to the target frame f_b; these are highlighted in cyan in the figure. As the reader can convince him-/herself, forward disocclusions are regions that are not visible in f_a, while reverse disocclusions are not visible in f_c, and should only be predicted from the reference frame they are visible in. Appropriate occlusion handling is particularly important on high-resolution content, where such regions can become quite large, in particular for fast moving objects.

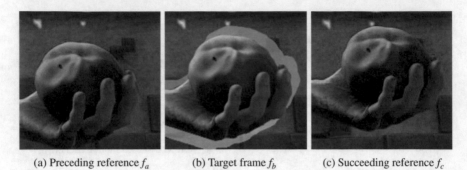

(a) Preceding reference f_a (b) Target frame f_b (c) Succeeding reference f_c

Fig. 3.4 Importance of occlusion-handling. Some regions in the target frame f_b are only visible in either of the reference frames f_a or f_c. In this example, yellow and cyan regions are not visible in f_a and f_c, respectively, and therefore should only be predicted from f_c and f_a, respectively

3.2.1   TFI Schemes with Target-Based Motion Anchoring

Schemes where the motion is anchored at the target frame are generally referred to as *bilateral schemes*; since we find this term ambiguous, we often use the more expressive term of *target-based* motion anchoring. While in a video compression scheme, this target-based motion anchoring seems the "natural" choice, it might be a bit more surprising how this can be achieved in a TFI scheme, where the target frame is not available. The underlying principle of bilateral schemes is as follows: the (non-existing) target frame is partitioned into blocks. For each block, the linear motion is searched for that results in the minimum block distortion between the corresponding regions in the two reference frames f_a and f_c, as illustrated in Fig. 3.3a. An example of a bilateral motion estimation scheme was proposed by Choi et al. [19]; in this method, block artefacts are reduced using an adaptive OBMC based on the reliability of neighbouring motion vectors. Wang et al. [20] perform motion-compensated prediction of the intermediate frame from both reference frames independently, and then blend these predictions together using a trilateral filter. Veselov and Gilmutdinov [21] propose a hierarchical bidirectional multi-stage motion estimation algorithm. They partition the target frame into non-overlapping, hierarchical blocks, and approximate the "true" motion flow. Each pixel is blended from multiple reference pixels. Raket et al. [22] perform a symmetric total-variation optical flow estimation at the unknown target frame, which is able to roughly halve the computation time compared to traditional bidirectional motion estimation schemes that estimate both forward and backward flows. Common to all target-anchored motion estimation schemes is the fact that disoccluded regions are not explicitly handled. Furthermore, new motion has to be estimated for every target frame that is to be interpolated, making such methods less attractive for large framerate upsampling factors.

3.2.2 TFI Schemes with Reference-Based Motion Anchoring

Reference-based motion anchoring schemes partition a reference frame (f_a or f_c) into blocks, and find for each block the motion vector MV that corresponds to the best match in the other reference frame (f_c or f_a). Then, each block is mapped to the target frame by appropriate scaling of the value of the MV. Jeong et al. [23] perform a multi-hypothesis motion estimation. The best motion hypothesis is selected by optimizing the cost function of a labelling problem. Pixels in the target frame are computed as a weighted combination of several pixels from the reference frame. They show improved reconstruction quality, at the expense of a significant increase in computational complexity. Dikbas et al. [18] use an adaptive interpolation between the forward and backward warped frame. Their method has low computational complexity, but the implicit occlusion handling can lead to visual distortions if disoccluded regions become large. Chin and Tsai [24] estimate a forward optical flow field using [25], and apply the motion to each pixel location. Holes and multiple mapped locations in the upsampled frame are handled using simple heuristics based on texture information.

3.2.3 Occlusion Handling

One might be led to believe that target-based TFI schemes have the advantage that, by design, there are no holes or double mappings to be resolved in the target frame. On the negative side, occluded regions cannot be explicitly handled, since they are not observed. In reference-based schemes, there will inevitably be regions in the target frame where multiple blocks overlap; this happens on the leading side of moving objects, for example when a foreground object moves over a region in the background. On the trailing side of objects in motion, there will be regions that are not hit by any block; the resulting holes have to be handled. While this might appear to be a disadvantage, it turns out that it also enables a better handling of such regions than the "opportunistic" handling of bilateral schemes, which inevitably blend foreground and background information, which results in ghosting artefacts.

Several reference-based frame interpolation methods have been proposed that explicitly handle occluded regions; such methods in general show improved performances compared to similar methods without occlusion handling. Kim et al. [26] estimate the forward and backward motion between the two reference frames, and then use linearity checking between the forward and backward flow to detect occluded regions. Cho et al. [27] use a bidirectional motion estimation scheme that is based on feature trajectory tracking, which allows the authors to detect occluded regions. Kim et al. [28] estimate motion using a *modified 3DRS (M3DRS)* in a spatial hierarchy with subsequent motion vector refinement using a temporal motion smoothness criterion. Motion vectors from adjacent blocks are grouped together, and overlapping regions are resolved by selecting the one with lower SAD between the two reference frames. Herbst et al. [29] perform bidirectional motion compensation by comput-

ing both a forward and a backward flow, which are then independently mapped to
the target frame. Assuming that each pixel location of the target frame is visible
in at least one reference frame, occluded regions are detected using a flow consistency criterion. Double mappings in the target frame are handled by assigning the
motion with larger velocity as foreground motion, which, as the authors point out,
is not always valid; in particular, the assumption that the faster moving object is the
foreground object fails if the background motion is larger than the motion of the
foreground object. Stich et al. [30] propose a "perception-motivated" frame interpolation method. They partition the reference frame into superpixels, which are then
mapped using homographies. As in [29], they resolve double mappings by assuming
that the faster moving pixel is closer to the camera. Occluded regions are detected
using a connectedness criterion proposed in [31]. Lu et al. [8] propose a *multiframe*
based method which identifies occluded regions as well as double mappings in the
upsampled frame using four reference frames. Since their method is block-based, the
frames are interpolated using adaptive OBMC to reduce blocking artefacts, which
also reduces the amount of high-frequency content of the upsampled frames.

In Fig. 3.5, we compare the interpolated frames produced by three state-of-the-art
TFI methods on a crop of a frame of the "Kimono2" sequence, where a woman walks
to the right. Figure 3.5d shows the result produced by Veselov et al. [21], which uses
a target-based anchoring, and hence is not able to handle occluded regions. One can

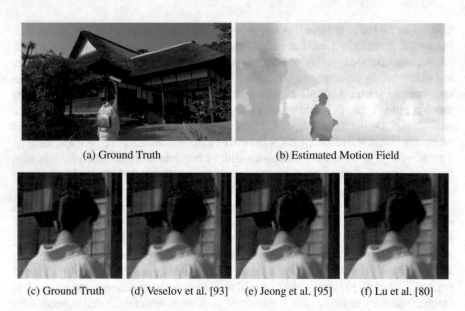

(a) Ground Truth   (b) Estimated Motion Field

(c) Ground Truth   (d) Veselov et al. [93]   (e) Jeong et al. [95]   (f) Lu et al. [80]

Fig. 3.5 Example interpolated frames from the "Kimono" sequence, obtained using different TFI
schemes. **a** shows the original frame; **b** shows the motion, estimated using MDP-flow [13]; the
second row shows crops of **c** the ground truth, **d** the interpolated frame using a target-anchored
scheme [21] *without* occlusion handling, as well as reference-anchored schemes with **e** *implicit*
[23] and **f** *explicit* [8] occlusion handling

see the ghosting artefacts around the head of the woman. In Fig. 3.5e, we show the interpolated frame produced by Jeong et al. [23], which does not explicitly handle occlusions; however, their method employs a involved texture optimization step, which is able to improve the results. Lastly, Fig. 3.5f shows the results of Lu et al. [8], which is a reference-based scheme with explicit occlusion handling.

The performance of a TFI scheme is normally evaluated by dropping all odd frames from a video sequence, and then interpolating the odd frames from the even ones. The interpolated frame $\hat{f}_b$ is then compared to the original f_b, usually in terms of PSNR. To the best of our knowledge, with the exception of [32, 33], existing TFI schemes interpolate frames under a constant velocity assumption, and hence PSNR comparisons are only really justified for sequences where the objects are following a constant velocity motion between any pair of temporally adjacent even frames. As stated in [32], the incorporation of higher-order motion models could have a significant impact on the compression performance; in Sect. 7.3, we experimentally show that this is indeed the case. One reason for constant velocity assumption is that in order to incorporate higher order motion models (e.g., acceleration, jerk,...), an accurate model of the underlying motion flow is required; such motion models are absent in many of the prior schemes. In [33], the experiments are performed on reasonably simple sequences with hardly any disocclusions.

3.2.4 Observations and Recommendations

Based on the literature review presented in this chapter, we now provide a list of observations pertaining to TFI methods.

- Bilateral TFI schemes, where the motion fields are estimated at the (non-existing) target frame, are unable to detect and hence handle occluded regions. Furthermore, motion has to be estimated for every frame to be interpolated in between the two reference frames, which does not scale well with higher frame upsampling factors;
- TFI schemes that employ block motion are unable to represent non-translational motion, such as zoom and rotation. Furthermore, block-based schemes usually employ OBMC to reduce blocking artefacts. This inevitably blurs textured regions, which "unnecessarily" reduces the quality of the interpolated frames;
- Only few TFI methods have been proposed that employ optical flow as opposed to block motion, mostly due to their relatively high computational complexity. However, we observe that the computational cost of block-based schemes that aim at high-quality interpolated frames is comparable to state-of-the-art optical flow methods;
- Many TFI papers evaluate their method on low-resolution, low-quality sequences, which are not reflective of today's typical video content. In particular, high-quality motion and occlusion handling become much more important. Evaluation of TFI performance should therefore be performed on high-quality content;

- Relatively few TFI methods explicitly handle regions around moving objects; out of these, most effort is on handling *disoccluded regions* (i.e., holes), which arise on the trailing side of moving objects. *Double mappings* that happen on the leading side of moving objects, are very rarely mentioned at all in the literature; if they are handled, the reasoning is either based on texture information, or selecting the largest velocity motion vector as foreground motion, which, in general, is not correct.

We conclude that the best TFI results can be expected from a scheme that uses "physical" rather than "block" motion, as can be obtained using state-of-the-art optical flow estimation methods. Furthermore, in order to be able to handle occluded regions and easily incorporate framerate upsampling factors larger than two without having to re-estimate motion, a reference-based motion anchoring should be used.

3.3 Summary

In this second background chapter, we introduced various true motion estimation schemes that aim at estimating the "true" trajectory of objects, or at least improve on the prediction field that is used in video compression schemes. For their ease of implementation and relatively low computational complexity, most TFI schemes employ block motion fields. In order to produce high quality results, the block motion is further refined using implicit and explicit smoothness constraints. This results in the fact that best-performing block-based TFI methods use sophisticated, time-consuming motion estimation schemes, with computational complexities similar to modern optical flow estimation schemes; however, optical flow schemes are able to estimate superior quality motion fields which do not suffer from artificial block artefacts. All motion anchoring strategies we propose in this thesis employ a reference-based motion anchoring with "physical" motion, and TFI is performed to form predictions of the target frames. With such a seamless integration of TFI with video compression, the computationally expensive motion (re-)estimation at the decoder can be avoided, which enables high-quality frame interpolation.

References

1. H. Schwarz, D. Marpe, T. Wiegand, Overview of the scalable video coding extension of the H.264/AVC standard. IEEE Trans. Circuit Syst. Video Technol. **17**(9), 1103–1120 (2007)
2. P. Helle, H. Lakshman, M. Siekmann, J. Stegemann, T. Hinz, H. Schwarz, D. Marpe, T. Wiegand, A scalable video coding extension of HEVC, in *Proceedings of the IEEE Data Compression Conference* (2013)
3. S.H. Chan, T.Q. Nguyen, LCD motion blur: modeling, analysis, and algorithm. IEEE Trans. Image Process. **20**(8), 2352–2365 (2011)
4. B. Girod, A.M. Aaron, S. Rane, D. Rebollo-Monedero, Distributed video coding. Proc. IEEE **93**(1), 71–83 (2005)

5. G. De Haan, P.W.A.C. Biezen, H. Huijgen, O.A. Ojo, True- motion estimation with 3-D recursive search block matching. IEEE Trans. Circuit Syst. Video Technol. **3**(5), 368–379 (1993)
6. A. Beric, G. De Haan, J. Van Meerbergen, R. Sethuraman, Towards an Efficient High Quality Picture-rate Up-converter, 2003
7. T. Ha, S. Lee, J. Kim, Motion compensated frame interpolation by new block-based motion estimation algorithm. IEEE Trans. Consum. Electron. **50**(2), 752–759 (2004)
8. Q. Lu, N. Xu, X. Fang, Motion-compensated frame interpolation with multiframe based occlusion handling. IEEE J. Disp. Technol. **11**(4), (2015)
9. B.K. Horn, B.G. Schunck, Determining optical flow. Artif. Intell. **17**, 185–203 (1981)
10. T. Brox, A. Bruhn, N. Papenberg, J. Weickert, High accuracy optical flow estimation based on a theory for warping, in *European Conference on Computer Vision* (2004), pp. 25–36
11. D. Sun, S. Roth, J. Lewis, M.J. Black, Learning optical flow, in *European Conference on Computer Vision* (2008), pp. 83–97
12. A. Wedel, D. Cremers, T. Pock, H. Bischof, Structure-and motion- adaptive regularization for high accuracy optic flow, in *Proceedings of the IEEE International Conference on Computer Vision* (2009), pp. 1663–1668
13. L. Xu, J. Jia, Y. Matsushita, Motion detail preserving optical flow estimation. IEEE Trans. Pattern Anal. Mach. Intell. **34**, 1744–1757 (2012)
14. J. Wulff, M.J. Black, Modeling blurred video with layers, in *European Conference on Computer Vision* (2014), vol. 8694, pp. 236–252
15. J. Revaud, P. Weinzaepfel, Z. Harchaoui, C. Schmid, EpicFlow: edge-preserving interpolation of correspondences for optical flow, in *Proceedings of the IEEE Conference on Computer Vision and Pattern Recognition* (2015)
16. M. Menze, C. Heipke, A. Geiger, Discrete optimization for optical flow, in *German Conference on Pattern Recognition* (2015)
17. Q. Chen, V. Koltun, Full flow: optical flow estimation by global optimization over regular grid, in *Proceedings of the IEEE Conference on Computer Vision and Pattern Recognition* (2016), vol. 2016, pp. 4706–4714
18. S. Dikbas, Y. Altunbasak, Novel true-motion estimation algorithm and its application to motion-compensated temporal frame interpolation. IEEE Trans. Image Process. **22**(8), 2931–2945 (2013)
19. B.-D. Choi, J.-W. Han, C.-S. Kim, S.-J. Ko, Motion-compensated frame interpolation using bilateral motion estimation and adaptive overlapped block motion compensation. IEEE Trans. Circuit Syst. Video Technol. **17**(4), 407–416 (2007)
20. C. Wang, L. Zhang, Y. He, Y.-P. Tan, Frame rate up-conversion using trilateral filtering. IEEE Trans. Circuit Syst. Video Technol. **20**(6), 886–893 (2010)
21. A. Veselov, M. Gilmutdinov, Iterative hierarchical true motion estimation for temporal frame interpolation, in *IEEE International Workshop on Multimedia Signal Processing* (2014)
22. L.L. Rakêt, L. Roholm, A. Bruhn, J. Weickert, Motion compensated frame interpolation with a symmetric optical flow constraint. Adv. Vis. Comput. 447–457 (2012)
23. S.-G. Jeong, C. Lee, C.-S. Kim, Motion-compensated frame interpolation based on multihypothesis motion estimation and texture optimization. IEEE Trans. Image Process. **22**, 4497–4509 (2013)
24. Y. Chin, C.-J. Tsai, Dense true motion field compensation for video coding, in *Proceedings of the IEEE International Conference on Image Processing* (2013), pp. 1958–1961
25. D. Sun, S. Roth, M.J. Black, Secrets of optical flow estimation and their principles, in *Proceedings of the IEEE Conference on Computer Vision and Pattern Recognition* (2010), pp. 2432–2439
26. D. Kim, H. Lim, H. Park, Iterative true motion estimation for motion-compensated frame interpolation. IEEE Trans. Circuit Syst. Video Technol. **23**(3), 445–454 (2013)
27. Y.-H. Cho, H.-Y. Lee, D.-S. Park, Temporal frame interpolation based on multiframe feature trajectory. IEEE Trans. Circuit Syst. Video Technol. **23**(12), 2105–2115 (2013)
28. D. Kim, H. Park, An efficient motion-compensated frame for high-resolution videos. IEEE J. Disp. Technol. **11**(7), 580–588 (2015)

29. E. Herbst, S. Seitz, S. Baker, Occlusion Reasoning for Temporal Interpolation using Optical Flow, Department of Computer Science and Engineering, University of Washington, Technical report. UW-CSE-09- 08-01, 2009
30. T. Stich, C. Linz, C. Wallraven, D. Cunningham, M. Magnor, Perception-motivated interpolation of image sequences. ACM Trans. Appl. Percept. **8**(2), 1–25 (2011)
31. W.R. Mark, L. McMillan, G. Bishop, Post-rendering 3D warping, in *Proceedings of the Symposium on Interactive 3D Graphics* (1997), pp. 7–16
32. R. Leonardi, A. Iocco, Time-varying motion estimation on a sequence of images. Multimed. Commun. Video Coding, 309–315 (1996)
33. P. Csillag, L. Boroczky, Estimation of accelerated motion for motion-compensated frame interpolation. Vis. Commun. Image Process. 604–614 (1996)

Chapter 4
Motion-Discontinuity-Aided Motion Field Operations

As we have seen in Chap. 2, existing video codecs anchor motion information at the frame that is to be predicted. In this thesis, we part from this conventional wisdom and investigate new motion anchoring strategies for highly scalable video compression, where motion fields are anchored at reference frames instead. As it turns out, this has a number of advantages over the traditional way of anchoring motion at reference frames. One of the main challenges using such a *reference-based* motion anchoring is that in order to be used for predicting the target frames, motion fields need to be warped to the target frames. This process leads to holes in disoccluding regions, as well as double mappings in the warped motion fields.

In this chapter, we describe the two elementary operations on motion fields, namely motion field *inversion* and motion *inference*, which enable the change of the motion field anchoring and perform a bidirectional prediction of the target frame. The proposed methods only use motion information and derived motion discontinuity information to disambiguate and handle problematic regions around moving objects.

We start by presenting how motion discontinuities can be mapped from one frame to another in Sect. 4.1; essentially, one has to find out which side of the motion discontinuity belongs to the foreground object, since motion discontinuities "travel" with the foreground object. In Sect. 4.2, we describe how motion fields can be *inverted*; in particular, we propose a procedure that guarantees a motion assignment for any location in the target frame, and show how motion discontinuities can be used to identify the foreground motion in regions that get double mapped. We evaluate the motion field inversion process in a unidirectional TFI scenario, which potentially is an interesting application in its own right.

However, the true potential of the reference-anchoring lies in its bidirectional prediction capabilities, for which we require a second motion field operation we call motion *inference*, which are presented in Sect. 4.3. In Sect. 4.4, the two motion field

© Springer Nature Singapore Pte Ltd. 2018
D. Rüfenacht, *Novel Motion Anchoring Strategies for Wavelet-Based Highly Scalable Video Compression*, Springer Theses,
https://doi.org/10.1007/978-981-10-8225-2_4

operations are evaluated in a bidirectional TFI setting using our BOA-TFI scheme,[1] where a comprehensive evaluation and comparison with other state-of-the-art TFI schemes highlights the advantages of the proposed scheme.

4.1 Warping of Motion Discontinuities

One key distinguishing feature of the proposed scheme is the use of *motion discontinuity information* to identify foreground/background moving objects; it is used during the *inversion* of motion fields to resolve double mappings in regions of motion field folding (see Sect. 4.2), as well as to extrapolate motion in disoccluded regions during the motion field *inference* process (see Sect. 4.3). As mentioned in Sect. 2.3.3, we use a highly scalable motion discontinuity representation using breakpoints to efficiently code the piecewise-smooth motion fields. In this and the next chapter, we use breakpoints as motion discontinuity representation for reasoning in the temporal domain. In Sect. 6.1, we present a more generic mechanism for representing motion discontinuities, which does not depend on the existence of breakpoints.

In the following, we present how breakpoints can be transferred from reference frames to the target frame we seek to interpolate. As described in Sect. 2.3.3.2, breakpoints lie on grid *arcs*, and can be connected to form discontinuity line segments. The underlying idea for mapping discontinuity line segments from reference to target frames is that motion discontinuities displace with the foreground object. Since the presence of a breakpoint necessarily implies that the motion on either side of it is significantly different, the aim is to identify the *foreground* motion by performing a *breakpoint compatibility check (BCC)* between the two reference frames f_a and f_c, and then warp *compatible* line segments to the target frame by appropriately scaling the identified foreground motion.

We now provide a more detailed description of the two steps of the temporal breakpoint warping procedure, which is visualized in Fig. 4.1. In the following, we focus on one line segment, formed by connecting two breaks belonging to the same cell; the same procedure is repeated for all cells. Let $B_{a,j}$, $j \in \{1,2\}$, be the two breakpoints that form the line segment l_a, and let $\mathbf{u}_{a \to c,j}^{H_p}$, $p \in \{1,2\}$, be the motion on side p of breakpoint $B_{a,j}$ that maps from frame f_a to frame f_c. Then, in step (1) of the temporal breakpoint warping procedure (see Fig. 4.1), breakpoints are mapped to f_c as follows:

$$B_{c,j}^{H_p} = B_{a,j} + \mathbf{u}_{a \to c,j}^{H_p}. \tag{4.1}$$

We denote the warped line segments as $l_c^{H_p}$. In order to determine which line segment $l_c^{H_p}$ lies closer to a motion discontinuity in frame f_c, we form *search line segments* $s_j = [B_{c,j}^{H_1}, B_{c,j}^{H_2}]$, and extend them on both sides by (at most) half the length of

---

[1]Initial results of the BOA-TFI scheme were presented in [1]; the comprehensive evaluation presented in this chapter has appeared in [2].

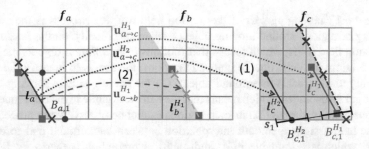

Fig. 4.1 Temporal induction of breakpoints, in order to warp motion discontinuity information from reference to target frames. (1) The discontinuity line segment l_a is mapped from f_a to frame f_c, using the motion on either side of the discontinuity. Search segments s are formed by connecting the endpoints of the line segments (e.g., $B_{c,1}^{H_1}$ and $B_{c,1}^{H_2}$); for one of the endpoints ($B_{c,1}^{H_1}$ in this example), the search segment intersects with motion discontinuities in f_c, which we call a *compatible* mapping. In step (2), the line segment is mapped to f_b, using the identified compatible (foreground) motion

s_j. The motion under hypothesis H_p that maps $B_{a,j}$ closer to the intersection of s_j with a line segment described by breakpoints in f_c (if any) is marked as *compatible*. If there is a hypothesis H_p for which $\mathbf{u}_{a \to c,j}^{H_p}$ is compatible for both j's (i.e., for both breakpoints that form the line segment), then $l_c^{H_p}$ is marked as *compatible line segment*.

In step (2) of the procedure (see Fig. 4.1), all compatible line segments are mapped to the target frame f_b using their compatible motion $\mathbf{u}_{a \to b,j}^{H_p}$. We note that in the absence of any other knowledge, $\mathbf{u}_{a \to b,j}^{H_p} = 0.5 \mathbf{u}_{a \to c,j}^{H_k}$. In the next section, we show how motion discontinuity information can be used to *invert* motion fields.

4.2 Motion Field Inversion

In this section, we describe how motion fields can be inverted, which is an essential operation in the reference-based anchoring schemes in order to map motion from reference to target frames, so that they can serve as prediction references. That is, from a motion field $M_{i \to j}$ which is anchored at frame f_i, pointing to frame f_j, we want to compute its inverse,

$$M_{j \to i} = (M_{i \to j})^{-1}, \tag{4.2}$$

which requires to establish a *one-to-one* mapping between locations in f_i and f_j. In regions around moving objects, regions might get uncovered, and hence they will never get mapped. For this reason, motion fields are not invertible in a mathematical sense. Nonetheless, the method we propose to warp motion fields is guaranteed to leave no holes, and hence enables the inversion of motion fields.

The most challenging part of the motion field warping process is the "correct" handling of regions around moving objects; more specifically, on the *leading* side, the motion field folds, which means that motions from *different* objects map to the same location. In such "double mapped" regions, the aim is to identify the motion belonging to the (local) foreground object. On the *trailing* side of moving objects, regions get disoccluded, which means that they are not visible in the reference frame. More specifically, the motion in such regions cannot be observed; the procedure we propose here assigns a smooth interpolation between background and foreground motion, which is reasonable in a unidirectional prediction scenario. In Sect. 4.3, where we consider a bidirectional prediction scenario, we will modify the motion in disoccluded regions. In the next section, we describe a method to warp motion fields that uses reasoning about motion discontinuities in order to handle traditionally difficult regions around moving objects.

4.2.1 Cellular Affine Warping of Motion

In this section, we describe a procedure that allows us to warp motion information from one frame to another. In order to account for expanding and contracting motion, we partition the available motion field $M_{i \to j}$ into small cells in the domain of the reference frame f_i, dividing each cell into two triangles, as shown in Fig. 4.2a. For the following discussion, we use a fixed size of 1×1, noting that the computational efficiency could be greatly improved by adopting larger cells in regions of smooth motion; we present a method to obtain such an adaptive mesh in Sect. 7.2.

We use $\mathbf{x} = (x, y)$ to indicate *continuous* locations, and $\mathbf{m} = [m, n]$ to indicate *discrete* (integer) locations. Since $M_{i \to j}$ only has motion at integer locations $\mathbf{m}$, we use $T_{i \to j}(\mathbf{x})$ to denote the *affine* interpolated motion of $M_{i \to j}$, which maps the continuous-valued location $\mathbf{x}_i$ from frame f_i to $\mathbf{x}_j$ in frame f_j. That is,

$$\mathbf{x}_j = \mathbf{x}_i + T_{i \to j}(\mathbf{x}_i). \tag{4.3}$$

In order to guarantee that the warped motion field from frame f_i completely covers the target frame f_j, we extend the affine flow field $T_{i \to j}$ by 1 pixel beyond the boundaries of frame f_i, assigning zero motion to the extension. In this way, the warped motion exhibits no holes. Due to its nature, we call the procedure of mapping triangles from one frame to another *cellular affine warping (CAW)*.

As we map triangles from f_i to f_j, whenever $T_{i \to j}(\mathbf{x}_i)$ falls onto an *integer* location $\mathbf{m}_j$ in the target frame f_j, we record its motion as

$$\hat{M}_{j \to i}[\mathbf{m}_j] = -T_{i \to j}(\mathbf{x}_i). \tag{4.4}$$

As illustrated in Fig. 4.3, one of three categories can be readily assigned to each mapped triangle: visible in both frames (black); disoccluded in target frame (yellow);

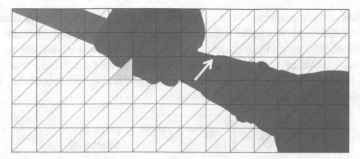

(a) Triangular partition of the reference motion field $M_{a \to c}$; we denote the continuous affine interpolated motion as $T_{a \to c}$.

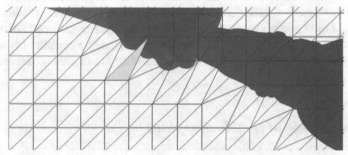

(b) Triangles are warped to the target frame f_b using $T_{a \to b}$, where they from a distorted mesh

Fig. 4.2 The proposed *cellular affine warping (CAW)* procedure partitions the reference *motion field* into triangles. Each such triangle is then mapped from the reference f_i to the target frame f_j, where each integer location gets assigned the corresponding affine motion. In regions that get disoccluded, triangles stretch without changing orientation (e.g., the yellow triangle), and the affine model assigns an interpolated value between the foreground and background motion, without leaving holes

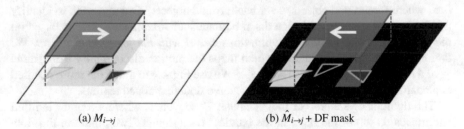

(a) $M_{i \to j}$ (b) $\hat{M}_{i \to j}$ + DF mask

Fig. 4.3 The proposed CAW method for inverting motion fields readily observes disocclusion (yellow) and folding (magenta) in the target frame f_j; the obtained *disocclusion and folding* (DF) mask will be valuable in a bidirectional prediction process

and folded (double mapping) in target frame (magenta). In other words, the motion inversion process allows us to obtain a disocclusion (and folding) mask without communicating any side information (see Fig. 4.3b); this is in stark contrast to video compression systems with a target-based motion anchoring, where this information has to be explicitly communicated as *side-information*. We record this valuable information in a disocclusion mask $\hat{S}_{j\rightarrow i}$:

$$\hat{S}_{j\rightarrow i}[\mathbf{m}] = \begin{cases} 0 & \mathbf{m} \text{ disoccluded in } (M_{i\rightarrow j})^{-1} \\ 1 & \text{otherwise} \end{cases}. \tag{4.5}$$

Each location in f_j will be assigned motion that can be used to predict f_j from f_i. Disoccluded regions, which arise on the *trailing* side of objects in motion, are assigned a smooth (stretched) motion field during the inversion process. On the leading side of moving objects, locations might be assigned multiple motion candidates due to *folding* of the triangular mesh. In the following section, we explain how reasoning about motion discontinuities can be used to identify the motion of the foreground object in such regions.

4.2.2   Resolving of Double Mappings

As the CAW procedure maps triangles from reference to target frames in order to compute the inverted motion field $\hat{M}_{j\rightarrow i}$, multiple triangles might overlap at location $\mathbf{m}_j$ in the target frame f_j. In other words, there are (at least) two locations $\mathbf{x}_i^1$ and $\mathbf{x}_i^2$ in f_i, which are mapped by $T_{i\rightarrow j}(\cdot)$ to the same location $\mathbf{m}_j$ in f_j, i.e.,

$$\mathbf{x}_i^1 + T_{i\rightarrow j}(\mathbf{x}_i^1) = \mathbf{x}_i^2 + T_{i\rightarrow j}(\mathbf{x}_i^2) = \mathbf{m}_j. \tag{4.6}$$

This happens on the *leading* side of moving objects, where the motion field is folding (i.e., where foreground objects cover background objects). Our approach to identify the foreground motion and hence disambiguate such *double mappings* is based on the observation that *motion discontinuities travel with the foreground object*. We therefore want to find the motion which maps the motion discontinuity from frame f_i to a motion discontinuity in frame f_j.[2] We use Fig. 4.4 to guide the more detailed explanation of the proposed method to resolve double mapped regions.

The figure shows a crop of the "Ambush 7" sequence, where a sceptre is lifted and moves on top of the snow in the (static)[3] background. Two points in Fig. 4.4a – one on the foreground object ($\mathbf{x}_i^1$), and one in the background ($\mathbf{x}_i^2$) – map to the same (integer) location $\mathbf{m}_j$ in the target frame f_j (Fig. 4.4b). The grey region in

---

[2]We remind the reader that the motion discontinuity information in the target frame f_j was mapped using the *hierarchical spatio-temporal breakpoint induction (HST-BPI)* procedure explained in Sect. 4.1.

[3]The example uses a static background for ease of explanation; we note that the method remains valid for moving backgrounds.

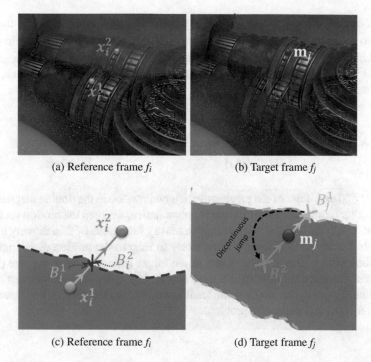

(a) Reference frame f_i (b) Target frame f_j

(c) Reference frame f_i (d) Target frame f_j

Fig. 4.4 Resolving of double mappings in the mapped motion field by reasoning about motion discontinuities (represented as red dashed lines around the sceptre). The key idea in identifying the foreground motion is that the motion discontinuities travel with the foreground object. In the example, $B_j^1 = B_i^1 + T_{i \to j}(B_i^1)$ maps closely to motion discontinuities, whereas $B_j^2 = B_i^2 + T_{i \to j}(B_i^2)$ is away from any motion discontinuity; consequently, $\hat{M}_{j \to i}[\mathbf{m}_j] = -T_{i \to j}(\mathbf{x}_i^1)$

Fig. 4.4d outlines the "true" inverse motion field at frame f_j, which is not known to the procedure; only the temporally induced breakpoints (green dashed line) are used to disambiguate the double mapping. The key idea behind the proposed method is that the two points $\mathbf{x}_i^1$ and $\mathbf{x}_i^2$ that create the double mapping in f_j must be separated in the reference frame f_i; moreover, along the line connecting the two points in f_i, denoted as l, there must be a discontinuity in the motion field.

If we map each point on the line l from frame f_i to frame f_j, using $T_{i \to j}$, we expect to see a discontinuous jump (red cross in Fig. 4.4c); we denote the points on either side of this discontinuous jump as B_i^1 and B_i^2. The location of these points mapped to f_j is

$$B_j^p = B_i^p + T_{i \to j}(B_i^p), \; p \in \{1, 2\}. \tag{4.7}$$

The importance of these points is that one of the B_j^ps is expected to fall close to a motion discontinuity, whereas the other one is not. The foreground motion is the motion of the point which maps closer to a motion discontinuity. In practice, we query the breakpoint cell structure for breaks, which can be done very efficiently.

That is, one B_j^p should fall into a cell that contains breakpoints, whereas the other falls into a cell without any breakpoints in the target frame f_j. Using C_j^p to denote the cell in frame f_j that the mapped breakpoint B_j^p falls into, we define a breakpoint *indicator* function $I_{breaks}(\cdot)$ that is 1 if the cell contains at least one break, and 0 otherwise. Then,

$$\hat{M}_{j \to i}[\mathbf{m}_j] = \begin{cases} -T_{i \to j}(\mathbf{x}_i^1) \ I_{breaks}(C_j^1) = 1 \wedge I_{breaks}(C_j^2) = 0 \\ -T_{i \to j}(\mathbf{x}_i^2) \ I_{breaks}(C_j^1) = 0 \wedge I_{breaks}(C_j^2) = 1 \ , \\ \hat{M}_{j \to i}^{old}[\mathbf{m}_j] \qquad\qquad\qquad\qquad\quad \text{otherwise} \end{cases} \qquad (4.8)$$

where $\hat{M}_{j \to i}^{old}[\mathbf{m}_j]$ denotes the previously assigned motion in the double mapped location. In other words, whenever the test is *inconclusive*, we keep the motion vector that was first assigned. The test is inconclusive mostly in regions of thin moving objects, where the background motion is more likely to map to the motion discontinuity of the trailing side of the moving object in the target frame. In Sect. 6.1, we present a different motion discontinuity measure that is able to distinguish between motion discontinuities on the trailing and the leading side of moving objects, and hence does not suffer from this particular problem.

4.2.3   Unidirectional Temporal Frame Interpolation

The ability to invert motion fields can be used to create motion fields that can predict the target frame f_b from the reference frame f_a. We evaluate the proposed motion inversion method on various examples of the Sintel sequence (see Sect. A.2), where ground truth motion is known between any pair of *consecutive* frames. In the following, let f_a and f_c denote two reference frames, and f_b be the target frame we wish to interpolate in between the two reference frames.

From the motion field $M_{a \to c}$, one can readily compute a *scaled* version that points to the intermediate frame f_b, as

$$\hat{M}_{a \to b} = \alpha M_{a \to c}, \qquad (4.9)$$

where $\alpha = 0.5$ in the case of doubling the framerate under a constant velocity assumption. In order to serve as prediction reference to interpolate frame f_b, $\hat{M}_{a \to b}$ is mapped to the target frame using the CAW procedure explained in Sect. 4.2.1, and double mappings are resolved as described in Sect. 4.2.2. Figures 4.5 and 4.6 show example inverted motion fields, as well as "unidirectionally" interpolated target frames $\hat{f}_b$, which are obtained by using the motion $\hat{M}_{b \to a}$ to map the texture information from the reference frame f_a to the target frame f_b.

One can see that the proposed motion inversion procedure is able to predict a credible interpolated frame in all regions *except* regions that get disoccluded between

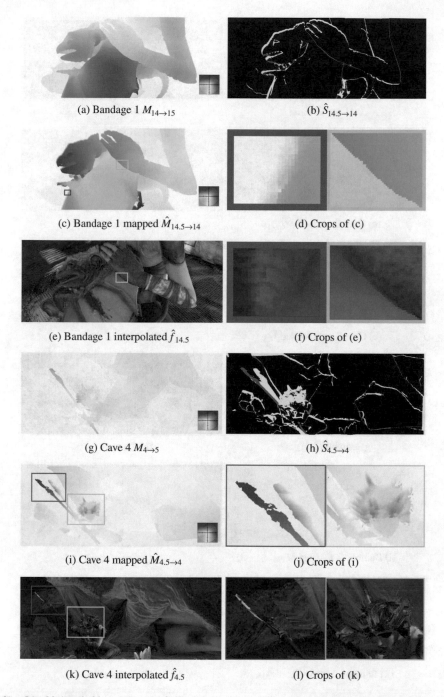

(a) Bandage 1 $M_{14\to15}$

(b) $\hat{S}_{14.5\to14}$

(c) Bandage 1 mapped $\hat{M}_{14.5\to14}$

(d) Crops of (c)

(e) Bandage 1 interpolated $\hat{f}_{14.5}$

(f) Crops of (e)

(g) Cave 4 $M_{4\to5}$

(h) $\hat{S}_{4.5\to4}$

(i) Cave 4 mapped $\hat{M}_{4.5\to4}$

(j) Crops of (i)

(k) Cave 4 interpolated $\hat{f}_{4.5}$

(l) Crops of (k)

Fig. 4.5 Motion field inversion results for the "Bandage 1" and "Cave 4" sequences from the Sintel dataset. **a/g** show the ground truth motion fields; **b/h** show the *estimated* disocclusion (yellow) and folding (magenta) masks; The second and fifth row show the mapped and *inverted* motion fields; the third and last row show the interpolated frames produced using the inverted motion fields

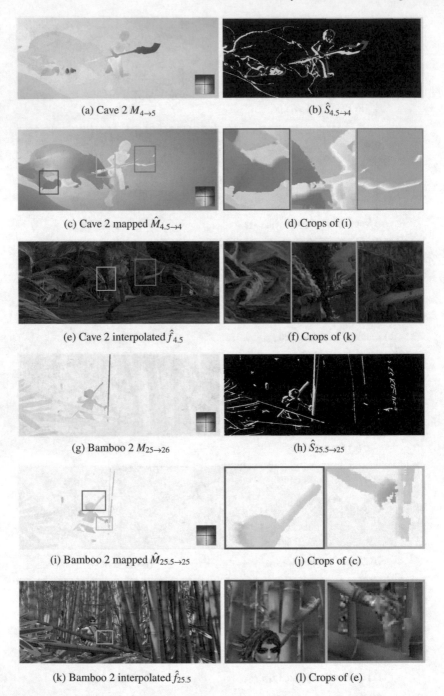

(a) Cave 2 $M_{4 \to 5}$

(b) $\hat{S}_{4.5 \to 4}$

(c) Cave 2 mapped $\hat{M}_{4.5 \to 4}$

(d) Crops of (i)

(e) Cave 2 interpolated $\hat{f}_{4.5}$

(f) Crops of (k)

(g) Bamboo 2 $M_{25 \to 26}$

(h) $\hat{S}_{25.5 \to 25}$

(i) Bamboo 2 mapped $\hat{M}_{25.5 \to 25}$

(j) Crops of (c)

(k) Bamboo 2 interpolated $\hat{f}_{25.5}$

(l) Crops of (e)

Fig. 4.6 Motion field inversion results for the "Cave 2" and "Bamboo 2" sequences from the Sintel dataset. **a/g** show the ground truth motion fields; **b/h** show the *estimated* disocclusion (yellow) and folding (magenta) masks; The second and fifth row show the mapped and inverted motion fields; the third and last row show the interpolated frames produced using the inverted motion fields

frames f_a and f_b (yellow regions in the disocclusion masks in the figures). In those regions, the CAW procedure assigns an interpolated motion between the motion of the foreground and the background object, which can hardly be seen as "physical" motion. Nonetheless, there is motion assigned at all locations. On the leading side of moving objects, where foreground objects move on top of background objects, double mappings are created (magenta regions in the disocclusion and folding masks of Figs. 4.5 and 4.6). One can see how in most regions, the correct foreground motion is identified.

In the next section, we present the second motion field operation, which will allow us to obtain a bidirectional prediction scheme with occlusion handling.

4.3 Reverse Inference of Motion Fields

Around moving object boundaries, there will be regions that get *disoccluded* (e.g., uncovered) from frame f_a to f_b; such regions are not visible in frame f_a. While not true in general, it is at least highly likely that such regions are visible in frame f_c, which is why we are interested in obtaining $M_{c \to b}$. One could be tempted to estimate $M_{c \to a}$, and then compute $M_{c \to b}$ as a scaled version of $M_{c \to a}$; that is, apply the exact same method described in the previous section "backwards" in time. To the best of our knowledge, this strategy is the one followed by all existing TFI schemes with bidirectional prediction, including TFI schemes that employ optical flow [3].

We avoid this strategy for two main reasons:

1. Motion estimation is the most time-consuming process of TFI *and* in modern video compression systems. Furthermore, in a (highly scalable) video coder, this would be mostly redundant information;
2. It is very likely that $M_{a \to c} \neq (M_{c \to a})^{-1}$, in particular around moving objects. Hence, their scaled versions will not be geometrically consistent in frame f_b (see Fig. 4.7).

In this thesis, we propose to instead *infer* $\hat{M}_{c \to b}$, anchored at frame f_c, from $M_{a \to c}$ and its *scaled* version $\hat{M}_{a \to b}$, as follows:

$$\hat{M}_{c \to b} = \hat{M}_{a \to b} \circ (M_{a \to c})^{-1}, \tag{4.10}$$

where $\circ$ denotes the *composition* operator. That is, each (integer) location $\mathbf{m}_c$ in $\hat{M}_{c \to b}$ gets assigned motion according to

$$\hat{M}_{b \to c}[\mathbf{m}_c] = -\underbrace{T_{a \to c}(\mathbf{x}_a)}_{=\frac{1}{\alpha}T_{a \to b}(\mathbf{x}_a)} + T_{a \to b}(\mathbf{x}_a) = \underbrace{\left(1 - \frac{1}{\alpha}\right)}_{<0 \ \forall \alpha \in]0,1[} T_{a \to b}(\mathbf{x}_a), \tag{4.11}$$

where $\mathbf{m}_c = \mathbf{x}_a + T_{a \to c}(\mathbf{x}_a)$, and α is the scaling factor. Because the inferred motion field is pointing in the "reverse" direction of the motion fields that are used to compose it, we refer to this form of motion inference as *reverse inference*.

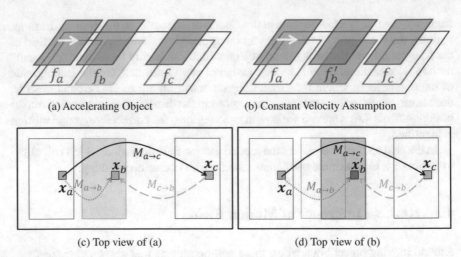

(a) Accelerating Object   (b) Constant Velocity Assumption

(c) Top view of (a)   (d) Top view of (b)

Fig. 4.7 Illustration of the concept of geometrical consistency. **a/c** show the true trajectory of the foreground object. **b/d** show how the inferred motion $M_{c \rightarrow b}$ follows the scaled motion $M_{a \rightarrow b}$, which means that they point to the same geometrical location in the target frame

The fact that $M_{c \rightarrow b}$ is completely defined by $M_{a \rightarrow c}$ and $M_{a \rightarrow b}$ has the key advantage that $M_{c \rightarrow b}$ always "follows" $M_{a \rightarrow b}$, such that the two motion fields involved in the prediction of frame f_b are *geometrically consistent*. This highly desirable property is illustrated in Fig. 4.7. In practice, this means that the predicted target frame will be significantly less blurred and contain less ghosting than traditional TFI approaches; examples are provided in Figs. 4.12 and 4.13.

By and large, the CAW procedure for motion field inference provides an excellent prediction for the "original" $M_{c \rightarrow b}$ field. However, in regions of f_c which correspond to background information that was occluded in f_a, the CAW procedure produces a poor prediction. As we have seen in the previous section, these *disoccluded* regions correspond to stretched triangles in frame f_c, where the CAW procedure infers a smooth transition between the background and foreground motions (see disoccluded regions in Figs. 4.5 and 4.6 for examples). For these regions, we propose to use a more realistic motion assignment based on *piecewise-constant motion extrapolation*, which we explain in the following section.

4.3.1 Discontinuity-Guided Background Motion Extrapolation in Disoccluding Triangles

The aim of the *reverse motion inference* process is to obtain a motion field $\hat{M}_{c \rightarrow b}$, anchored at frame f_c, and pointing to f_b, which is as close to a "true" motion field as possible. However, the affine interpolated motion assigned by the CAW procedure

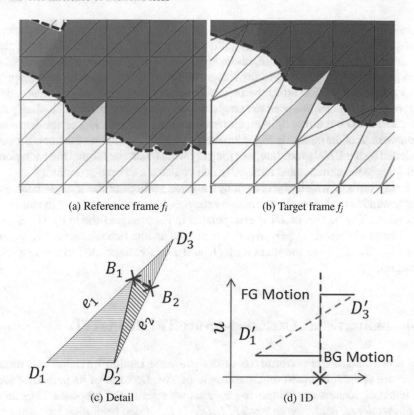

(a) Reference frame f_i (b) Target frame f_j

(c) Detail   (d) 1D

Fig. 4.8 Close-up of the scene in Fig. 4.2, to illustrate the motion extrapolation technique applied in disoccluded regions. **a** shows a triangle in the reference frame f_i, which straddles a motion discontinuity boundary. **b** shows the warped, stretched triangle in the target frame f_j. Instead of linearly interpolating motion from foreground to background, we instead extrapolate motion from the vertices to the motion discontinuity boundary, represented by B_1 and B_2; this results in sharp boundaries, as exemplified in (**d**), where the blue dotted line corresponds to linearly interpolated motion, and the grey solid line corresponds to extrapolated motion

in disoccluded regions can hardly be seen as true motion. In this section, we show how more meaningful motion can be assigned in regions of disocclusions.

In the absence of new motion appearing in regions that get disoccluded between frames f_a and f_c, a good estimate for the motion in such regions can be obtained by extrapolating the motion of the respective triangle vertices to motion discontinuity boundaries. For most of the disoccluded triangle, this means that background motion should be extrapolated; only a small (if any) part of the triangle covers the foreground object. We use Fig. 4.8 to explain the details of the proposed background motion extrapolation technique.

Whenever a triangle is stretching as it is mapped from a reference to a target frame, we expect it to intersect with motion discontinuities in the target frame; this is because some of its vertices belong to the background (possibly in motion), and some belong to the (possibly moving) foreground object. In Fig. 4.8c, D_1' and D_2'

belong to the background object, whereas D_3' belongs to the foreground. The warped triangle has two edges that intersect with motion discontinuities, which we denote as e_1 and e_2. As mentioned before, instead of interpolating a value transitioning from background (D_1' in Fig. 4.8) to the foreground motion D_3', we want to extrapolate the background motion up to the motion boundary; likewise, on the other side of the motion boundary, we want to extrapolate the foreground motion. To clarify this, we show a 1D cut along e_1, formed by connecting D_1' and D_3', of the *horizontal component* $\boldsymbol{u}$ of the motion in Fig. 4.8d; the dashed blue line shows the smooth motion assigned by the CAW procedure, and the gray solid (staircase) shows the background and foreground extrapolated motion, which contains a sharp discontinuity.

Irrespective of which object each of the three vertices of the triangle belongs to (foreground or background), the motion extrapolation method performs the following procedure: The motion of D_3' is extrapolated in the triangle formed by D_1', B_1, and B_2. The quadrilateral (D_1', D_2', B_1, B_2), is broken up into two triangles (D_1', D_2', B_1), and (D_2', B_1, B_2), and the motion of D_1' and D_2' is extrapolated in the respective triangles.

4.4   Bidirectional, Occlusion-Aware TFI (BOA-TFI)

We now evaluate the performance of the proposed motion inversion and motion inference operations based on the example of TFI. Because of its particular focus on handling regions around moving objects, we refer to the proposed TFI scheme as *bidirectional, occlusion-aware TFI (BOA-TFI)*.[4] Video resolution has seen a significant increase in recent years, while the framerate has not dramatically changed; what this means is that the expected size of disoccluded regions is larger, which makes appropriate handling of such regions more important. By contrast, the handling of occluded regions on low-resolution video (e.g., CIF and lower) is of smaller importance, since they tend to be small. On such low-resolution sequences, our TFI method performs similarly to existing TFI methods, and sometimes even worse, because we do not apply any texture optimizations to our interpolated frames. As part of this thesis, we want to highlight the importance of better motion and interpolation methods for high-resolution data; for this reason, all experiments are performed on high-resolution video sequences.

4.4.1   Method Overview

Guided by Fig. 4.9, we now give an overview of the proposed BOA-TFI method, for the case of doubling the framerate; we note, however, that the proposed scheme can readily accommodate frame upsampling factors larger than 2. We remind the reader

---

[4]This work was published in [1, 2].

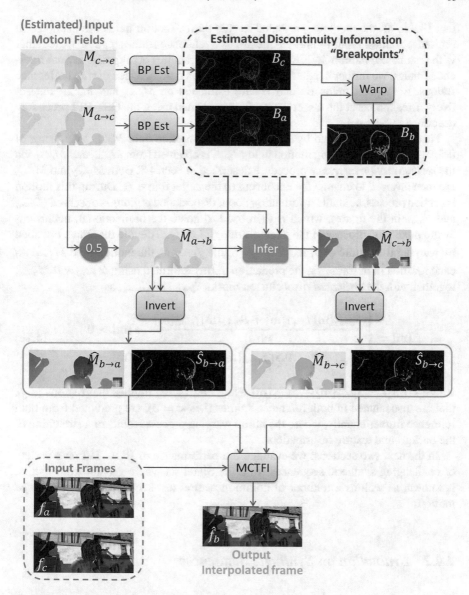

Fig. 4.9 Overview of the proposed BOA-TFI method. In addition to the two reference frames f_a and f_c, the inputs to the scheme are a (potentially estimated) motion field $M_{a \to c}$, as well as breakpoint fields *estimated* on $M_{a \to c}$ for frame f_a, and on $M_{c \to e}$ (only used to obtain breakpoints) for frame f_c. In the first step, estimated breakpoints at reference frames f_a and f_c (B_a and B_c) are transferred to the target frame f_b (B_b). Next, $M_{a \to b}$ is obtained by halving its parent motion field $M_{a \to c}$. $M_{a \to c}$ and $\hat{M}_{a \to b}$ are then used to *infer* the motion field $\hat{M}_{c \to b}$. The last step consists of inverting $\hat{M}_{a \to b}$ and $\hat{M}_{c \to b}$ to obtain $\hat{M}_{b \to a}$ and $\hat{M}_{b \to c}$. During the motion inversion process, we compute disocclusion masks $\hat{S}_{b \to a}$ and $\hat{S}_{b \to c}$, which are used to guide the bidirectional *motion-compensated temporal frame interpolation (MCTFI)* process to temporally interpolate the frame $\hat{f}_b$. Breakpoints are used to resolve double mappings and handle occluded regions during both the motion *inference* and *inversion* process

that like in all the methods presented in this thesis, motion fields are anchored at reference frames. Input to the scheme are two reference frames f_a and f_c, together with (estimated) motion $M_{a \to c}$. Since the scheme requires breakpoints at *both* reference frames, we further need $M_{c \to e}$, i.e., the motion field between the next reference frames. Breakpoint fields B_a and B_c are estimated on $M_{a \to c}$ and $M_{c \to e}$, respectively. Breakpoints at the target frame f_b are obtained using the HST-BPI procedure described in Sect. 4.1.

Next, $\hat{M}_{a \to b}$ is obtained by *scaling* the parent motion field $M_{a \to c}$ by a factor of 0.5. The backward pointing motion field $\hat{M}_{c \to b}$ is obtained from $M_{a \to c}$ and $\hat{M}_{a \to b}$ via the *reverse motion inference* procedure described in Sect. 4.3. Both $\hat{M}_{a \to b}$ and $\hat{M}_{c \to b}$ are then *inverted* to change the anchoring to the target frame f_b. During this motion inversion process, valuable information about disoccluded regions is observed ($\hat{S}_{b \to a}$ and $\hat{S}_{b \to c}$ in the figure), which is then used to guide the bidirectional, occlusion-aware prediction process of the target frame $\hat{f}_b$. Let $f_{i \to j}$ denote the frame obtained by warping the texture of f_i to the target frame f_j, using the motion field $\hat{M}_{j \to i}$. At each location $\mathbf{m}$ in frame f_b, the prediction $\hat{f}_b[\mathbf{m}]$ is formed using $\hat{M}_{b \to a}$ and $\hat{M}_{b \to c}$, together with the *estimated* disocclusion masks $\hat{S}_{b \to a}$ and $\hat{S}_{b \to c}$, as

$$
\hat{f}_b[\mathbf{m}] = \begin{cases} \dfrac{\hat{S}_{b \to a}[\mathbf{m}] f_{a \to b}[\mathbf{m}] + \hat{S}_{b \to c}[\mathbf{m}] f_{c \to b}[\mathbf{m}]}{\kappa[\mathbf{m}]} & \kappa[\mathbf{m}] > 0 \\ 0.5 \big(f_{a \to b}[\mathbf{m}] + f_{c \to b}[\mathbf{m}] \big) & \kappa[\mathbf{m}] = 0 \end{cases}, \qquad (4.12)
$$

where $\kappa([\mathbf{m}]) = \hat{S}_{b \to a}[\mathbf{m}] + \hat{S}_{b \to c}[\mathbf{m}]$. Note how (4.12) implies that regions in $\hat{f}_b$ that are disoccluded in both reference frames (i.e., $\kappa = 0$), are predicted from both reference frames equally, where the affine warping process results in a stretching of the background texture information.

In the next two sections, we evaluate the performance of BOA-TFI on a variety of challenging synthetic sequences from the Sintel set where ground truth motion is known, as well as a number of common natural test sequences with *estimated* motion.

4.4.2 Evaluation on Synthetic Sequences

The focus of this chapter is on the motion inference process which produces geometrically consistent interpolated frames. For this to work, we need piecewise-smooth motion fields with discontinuities at moving object boundaries. In this section, in order to focus on the frame interpolation quality of BOA-TFI, we use challenging synthetic sequences from the Sintel dataset, where ground truth motion fields between adjacent frames are known. We provide results on natural sequences with *estimated* motion in the next section.

Figures 4.10 and 4.11 show example interpolated frames generated by BOA-TFI; full-resolution versions of the results, including animated versions, can be found on the website dedicated to the corresponding journal publication [2].[5] Figs. 4.10a/g and 4.11a/g show the ground truth motion fields, which contain a variety of types of motion such as translation, rotation, zoom, and panning; furthermore, the motion magnitudes are much larger than on most "common" natural sequences, resulting in large regions of disocclusion around moving objects, as visualized in the disocclusion masks in Figs. 4.10b/h and 4.11b/h. Because the ground truth motion fields for the Sintel sequence are only between adjacent frames, the frame we interpolate does not exist in the sequence, and hence we cannot compute a PSNR. However, what ultimately counts is the perceived quality, and in particular in scenes with accelerated motion, PSNR is a poor predictor of perceived quality.

One can see how the scheme is able to create high quality reconstructed frames. The crops in the third and the last row of Figs. 4.10 and 4.11 highlight difficult regions around moving object boundaries, where BOA-TFI switches from bidirectional to unidirectional prediction without smoothing the texture. Ideally, in the case of upsampling by a factor of 2 under a constant velocity assumption, as we assume in this experiment, $\hat{M}_{b \to a} = -\hat{M}_{b \to c}$. One can see that this is not always the case, especially in regions of complex motion and scene geometry, such as the sword in Fig. 4.10i/j, as well as the stick in Fig. 4.11i/j, which is wrongly assigned background motion in $\hat{M}_{b \to c}$. The reason for this is that in order to obtain $\hat{M}_{b \to c}$, two motion field inversion operations have to be performed, which are independent to the inversion required to obtain $\hat{M}_{b \to a}$. In Chap. 6, we will present a modification of the motion field anchoring presented here, where only one motion field inversion has to be performed; as we will see, this will guarantee that $\hat{M}_{b \to a} = -\hat{M}_{b \to c}$.

It is worth highlighting that the scheme presented in this chapter does not perform any texture optimization. In particular, the transition from uni- to bidirectional prediction can cause artefacts at the transition boundary if there are significant changes in illumination between the two reference frames. This can be observed in the right crop of the "Bandage 1" sequence (see Fig. 4.10f), and is most visible in the upper left part (i.e., the part of the wing which moves under the hand) which is only predicted from the left reference frame. The wing is significantly brighter in the left reference frame, and hence the bidirectionally predicted part of the wing is darker than the unidirectionally predicted part. In Sect. 6.3, we present ways of selectively optimizing the texture in such regions, which has a positive impact both on the objective and subjective quality of the interpolated frames. However, even without any texture optimizations, BOA-TFI is able to produce high quality interpolated frames. In the next section, we further provide results on a variety of common natural test sequences with estimated motion, where we show highly competitive results with state-of-the-art TFI methods.

---

[5]http://ivmp.unsw.edu.au/~dominicr/atsip_boa_tfi.html.

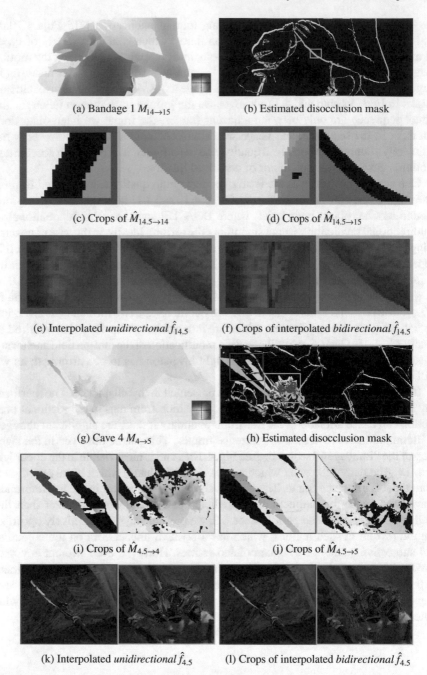

(a) Bandage 1 $M_{14 \to 15}$ (b) Estimated disocclusion mask

(c) Crops of $\hat{M}_{14.5 \to 14}$ (d) Crops of $\hat{M}_{14.5 \to 15}$

(e) Interpolated *unidirectional* $\hat{f}_{14.5}$ (f) Crops of interpolated *bidirectional* $\hat{f}_{14.5}$

(g) Cave 4 $M_{4 \to 5}$ (h) Estimated disocclusion mask

(i) Crops of $\hat{M}_{4.5 \to 4}$ (j) Crops of $\hat{M}_{4.5 \to 5}$

(k) Interpolated *unidirectional* $\hat{f}_{4.5}$ (l) Crops of interpolated *bidirectional* $\hat{f}_{4.5}$

Fig. 4.10 First set of TFI results on frames from the Sintel dataset. **a/g** show the ground truth motion fields; **b/h** show the *union* of the forward (yellow) and reverse (cyan) disocclusion masks produced by BOA-TFI; **c/i** and **d/j** show crops of the *estimated* motion fields $\hat{M}_{b \to a}$ and $\hat{M}_{b \to c}$, respectively, where black regions indicate disoccluded regions; **e/k** and **f/l** show crops of the TFI results obtained by unidirectional prediction and the proposed BOA-TFI method, respectively

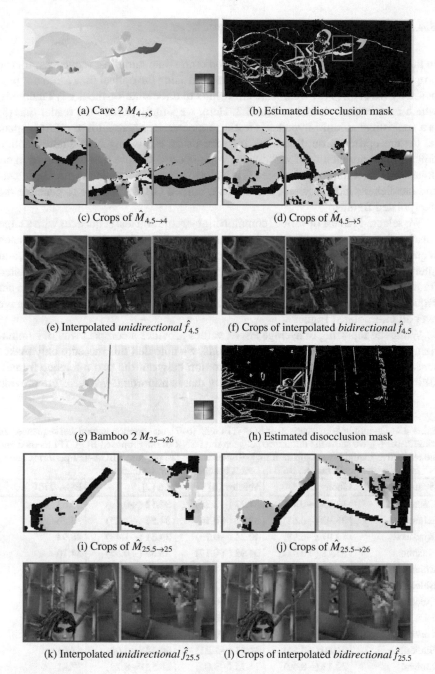

(a) Cave 2 $M_{4\to5}$

(b) Estimated disocclusion mask

(c) Crops of $\hat{M}_{4.5\to4}$

(d) Crops of $\hat{M}_{4.5\to5}$

(e) Interpolated *unidirectional* $\hat{f}_{4.5}$

(f) Crops of interpolated *bidirectional* $\hat{f}_{4.5}$

(g) Bamboo 2 $M_{25\to26}$

(h) Estimated disocclusion mask

(i) Crops of $\hat{M}_{25.5\to25}$

(j) Crops of $\hat{M}_{25.5\to26}$

(k) Interpolated *unidirectional* $\hat{f}_{25.5}$

(l) Crops of interpolated *bidirectional* $\hat{f}_{25.5}$

Fig. 4.11 Second set of TFI results on frames from the Sintel dataset. **a/g** show the ground truth motion fields; **b/h** show the *union* of the forward (yellow) and reverse (cyan) disocclusion masks produced by BOA-TFI; **c/i** and **d/j** show crops of the *estimated* motion fields $\hat{M}_{b\to a}$ and $\hat{M}_{b\to c}$, respectively, where black regions indicate disoccluded regions; **e/k** and **f/l** show crops of the TFI results obtained by unidirectional prediction and the proposed BOA-TFI method, respectively

4.4.3 Evaluation on Natural Sequences

In this section, we show results obtained on common natural test sequences; for the proposed BOA-TFI, motion fields are estimated using the optical flow estimator proposed by Xu et al. [4]. We compare our results to three state-of-the-art TFI methods, which have been reviewed in Sect. 3.2. Here, we simply remind the reader that [5] is a sophisticated multi-hypothesis testing framework, where a lot of effort is spent on texture optimization. Reference [6] focuses on estimating high-quality motion fields, which are then used without any sophisticated texture optimization to interpolate the target frame. Reference [7] is a block-based scheme that explicitly detects and handles occluded regions, and uses a modified OBMC scheme to generate the interpolated frames.

We selected 12 sets of various common high-resolution test sequences with a large variety of motion and texture complexity; in Sect. A.3, we show the first frame of each sequence. For each such sequence, we choose 11 adjacent *even* numbered frames, and interpolate the *odd* numbered frames in between them; this results in 10 interpolated frames per sequence, for a total of 120 interpolated frames. Table 4.1 presents the per sequence results, averaged over the 10 frames, as well as the performance average over all interpolated frames.

While the reporting of average PSNR values provides a compact way of summarizing the performance of the tested methods, we note that this measure only makes sense in regions where there is no acceleration between the two reference frames. Ultimately, it is the perceived visual quality that is important. We therefore provide

Table 4.1 Quantitative comparison of BOA-TFI with [5–7], on common natural test sequences. In parantheses $(\cdot)$, we show the difference between the PSNR of the proposed BOA-TFI method and the respective method we compare it to; "$-$" means that the proposed BOA-TFI performs better, "$+$" means worse performance. **Bold** indicates best per-row performance

Sequence	Jeong [5]	Veselov [6]	Lu [7]	BOA-TFI[†]
Cactus	33.15 ($-$0.49)	31.27 ($-$2.36)	**34.12** ($+$0.49)	33.63
Kimono1	33.93 ($+$0.68)	33.40 ($+$0.14)	**34.51** ($+$1.25)	33.26
Kimono2	39.97 ($-$0.98)	40.21 ($-$0.73)	39.51 ($-$1.43)	**40.94**
Rushhour	35.18 ($+$0.42)	34.93 ($+$0.17)	**35.30** ($+$0.55)	34.76
Shields1	35.90 ($-$0.66)	35.10 ($-$1.45)	35.89 ($-$0.66)	**36.55**
Shields2	33.87 ($-$3.89)	35.58 ($-$2.18)	33.52 ($-$4.24)	**37.76**
Stockholm	36.59 ($-$1.25)	37.12 ($-$0.72)	35.85 ($-$1.99)	**37.84**
Park	38.29 ($-$1.22)	38.84 ($-$0.67)	38.74 ($-$0.77)	**39.51**
Parkrun	30.63 ($-$1.17)	30.97 ($-$0.82)	30.50 ($-$1.29)	**31.79**
Station2	41.10 ($-$2.51)	41.41 ($-$2.21)	40.54 ($-$3.08)	**43.61**
Mobcal	29.13 ($-$8.68)	34.75 ($-$3.06)	29.53 ($-$8.28)	**37.81**
Terrace	33.29 ($-$4.34)	34.22 ($-$3.40)	33.66 ($-$3.96)	**37.62**
Average	35.08 ($-$2.01)	35.65 ($-$1.44)	35.14 ($-$1.95)	**37.09**

[†]Motion fields estimated using MDP [4]

(a) Parkrun f_{151}

(b) Ground truth Parkrun f_{151}

(c) BOA-TFI Parkrun $\hat{f}_{151}$

(d) Veselov *et al.* [93] Parkrun $\hat{f}_{151}$

(e) Jeong *et al.* [95] Parkrun $\hat{f}_{151}$

(f) Lu *et al.* [80] Parkrun $\hat{f}_{151}$

Fig. 4.12 Qualitative comparison of the proposed BOA-TFI scheme with state-of-the-art TFI methods on a frame of the "Parkrun" sequence. **a** shows the ground truth frame; **b–f** show crops of the ground truth, as well as interpolated frames produced by BOA-TFI [2], Veselov et al. [6], Jeong et al. [5], and Lu et al. [7], respectively

qualitative results for two of the sequences in Figs. 4.12 and 4.13. First off, all three TFI methods chosen for comparison are able to provide high quality interpolated frames for most of the sequences, in particular in regions *within* moving objects (i.e., away from moving object boundaries). The differences in PSNR values and visual quality are governed by *two major factors*:

How Regions of Global Motion are Interpolated Block-based methods usually employ a variant of OBMC, which tends to oversmooth the interpolated frames, resulting in significant blurring of the overall texture. In Fig. 4.12, this can be seen in highly textured regions such as the running man with the umbrella, as well as the

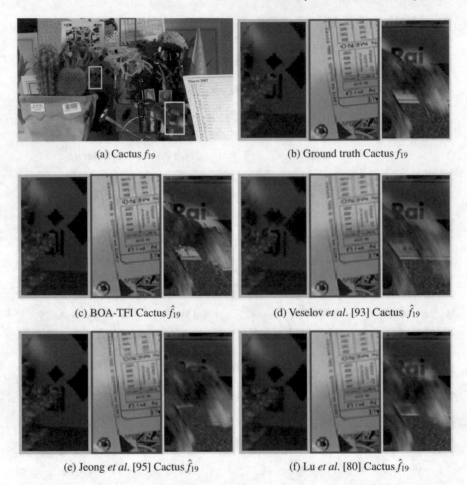

(a) Cactus f_{19} (b) Ground truth Cactus f_{19}

(c) BOA-TFI Cactus $\hat{f}_{19}$ (d) Veselov *et al.* [93] Cactus $\hat{f}_{19}$

(e) Jeong *et al.* [95] Cactus $\hat{f}_{19}$ (f) Lu *et al.* [80] Cactus $\hat{f}_{19}$

Fig. 4.13 Qualitative comparison of the proposed BOA-TFI scheme with state-of-the-art TFI methods on a frame of the "Cactus" sequence. **a** shows the ground truth frame; **b-f** show crops of the ground truth, as well as interpolated frames produced by BOA-TFI [2], Veselov et al. [6], Jeong et al. [5], and Lu et al. [7], respectively

text on the card of the Cactus sequence in Fig. 4.13. Lu et al.'s method [7] seems to be particularly affected by this, which can explain the significant drop in PSNR this method exhibits in some of the sequences reported in Table 4.1.

Occlusion-Handling Regions around moving objects are only visible from one reference frame, and hence should only be predicted from the frame in which they are visible. A proper occlusion handling can only be achieved if such regions are detected by the TFI algorithm. Of the three TFI schemes tested, only Lu et al. [7] explicitly handles occluded regions. The quality of the proposed occlusion handling is most visible in the "Parkrun" sequence around the left leg of the running man, where all other methods resort to frame averaging, while BOA-TFI creates a credible interpolation of the leg without any ghosting. On the "Cactus" sequence, the "10" (cyan crop) is properly interpolated by the proposed method.

In the current implementation of the proposed method, we do not perform any texture optimization. In regions which are highly affected by motion blur, such as the tiger in the "Cactus" sequence, this can create artificial high frequencies around moving object boundaries. As we will show in Sect. 6.3, the above-mentioned problems can be addressed in quite an elegant way by selectively smoothing the prediction in regions where there is a transition from uni- to bidirectional prediction.

4.5 Summary

In this chapter, we presented the two main motion field operations that will be used throughout this thesis. First, we proposed motion discontinuity aided motion *inversion* operation, which can be used to map motion fields from one frame to another. A key insight is that this motion field inversion process allows us to readily *observe regions* that get *disoccluded* in the target frame. Disoccluded regions are regions that are not visible in the respective reference frame. However, it is quite likely that such regions are visible in another reference frame. The second operation on motion fields, which we refer to as motion *inference*, is used to compose a motion field that relates the other reference frame with the target frame in a geometrically consistent way. Perhaps surprisingly, this "backward" pointing motion field is inferred from two forward pointing motion fields. The two motion field operations are evaluated in a TFI scenario on a large variety of both challenging synthetic sequences, as well as common natural sequences, where we show superior performance compared to three state-of-the-art TFI schemes.

The *motion centric* approach to TFI is well adapted to scalable compression schemes, because it allows the motion to be understood as part of a transform that is applied to the frame data; the proposed TFI scheme can then be understood as the inverse transform that would result if high temporal frequency details were omitted. In the next chapter, we present a highly scalable video compression scheme which has BOA-TFI as the main building block.

References

1. D. Rüfenacht, R. Mathew, D. Taubman, Bidirectional, occlusion-aware temporal frame interpolation in a highly scalable video setting, *Pict. Cod. Symp.* (2015)
2. D. Rüfenacht, R. Mathew, D. Taubman, Occlusion-aware temporal frame interpolation in a highly scalable video coding setting. APSIPA Trans. Signal Inf. Proc. **5** (2016)
3. F. Herbst, S. Seitz, S. Baker, Occlusion Reasoning for Temporal Interpolation using Optical Flow. Department of Computer Science and Engineering, University of Washington, Technical report UW-CSE-09- 08-01, 2009

4. L. Xu, J. Jia, Y. Matsushita, Motion detail preserving optical flow estimation. IEEE Trans. Patt. Anal. Mach. Intell. 1744–1757 (2012)
5. S.-G. Jeong, C. Lee, C.-S. Kim, Motion-compensated frame interpolation based on multihypothesis motion estimation and texture optimization. IEEE Trans. Image Proc. 4497–4509 (2013)
6. A. Veselov, M. Gilmutdinov, Iterative hierarchical true motion estimation for temporal frame interpolation, *IEEE International Workshop on Multimedia Signal Processing* (2014)
7. Q. Lu, N. Xu, X. Fang, Motion-compensated frame interpolation with multiframe based occlusion handling. IEEE J. Disp. Technol. **11**(4) (2015)

Chapter 5
Bidirectional Hierarchical Anchoring (BIHA) of Motion

In the previous chapter, we introduced two motion field operations that can be used to perform a bidirectional interpolation of a target frame from only two reference frames, together with the motion linking the reference frames; we refer to this scheme as BOA-TFI. In this chapter, we propose a *bidirectional hierarchical anchoring (BIHA)* of motion for highly scalable video compression, which employs BOA-TFI as a fundamental building block.[1] In Sect. 5.1, we present the BIHA scheme, and contrast it with the conventional anchoring of motion at target frames, as employed by all standardized video codecs. In Sect. 5.2, we detail the modifications to the motion-compensated temporal filtering *motion-compensated temporal filtering (MCTF)* process; in particular, we add an *update* step, which is beneficial in a video compression scenario.

The utility of the proposed scheme increases with the number of temporal decompositions. Section 5.3 presents how to scalably allocate the rate for all the different spatio-temporal subbands of texture, motion, and breakpoints. To further improve the R-D performance of the proposed highly scalable video compression scheme, we present in Sect. 5.4 an extension to the temporal breakpoint warping procedure (see Sect. 4.1); the resulting HST-BPI can account for the presence of partial breakpoint fields at target frames, and proves particularly useful at medium to low bit-rates. In Sect. 5.5, we compare the R-D performance of the BIHA scheme with the traditional motion field anchoring at target frames. We further show comparisons with SHVC, the latest standardized scalable video codec, where the proposed scheme shows promising results.

---

[1]We presented the initial idea of this work in [1, 2], which was extended with an analytical model for the scalable rate allocation as well as more comprehensive evaluation in [3].

© Springer Nature Singapore Pte Ltd. 2018
D. Rüfenacht, *Novel Motion Anchoring Strategies for Wavelet-Based Highly Scalable Video Compression*, Springer Theses,
https://doi.org/10.1007/978-981-10-8225-2_5

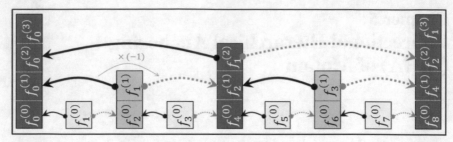

(a) Traditional anchoring of motion fields at target frames.

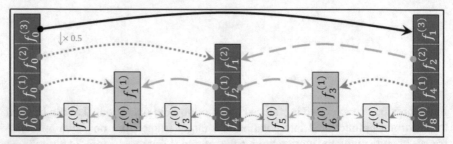

(b) Proposed anchoring of motion fields at the reference frames.

Fig. 5.1 Two ways of anchoring motion fields in a temporal hierarchy. We use $f_k^{(t)}$ to indicate the kth frame at level t of the temporal hierarchy. $\downarrow \times 0.5/(-1)$ indicates the scaling factor applied between motion fields. **a** Traditional anchoring as used in current video compression schemes, where the motion description is attached to the frame that is to be predicted; **b** The proposed *bidirectional hierarchical anchoring* (BIHA), where motion is described at reference frames instead, which are then mapped to the target frames in order to serve as prediction references. Each arrow indicates a *coded* motion field; solid arrows are *fully coded* motion fields, dotted arrows are *scaled* motion fields, and dashed arrows indicate *inferred* motion fields. One can see that in the traditional anchoring, motion information is only used for the prediction of one frame, whereas in the proposed BIHA scheme, motion information can be *scaled* from coarse to fine temporal levels, which is highly beneficial for scalable compression systems. In addition, roughly half the motion fields in the BIHA scheme are inferred, which, as we shall see, are very cheap to code

5.1 BIHA of Motion Fields for Highly Scalable Video Compression

Current state-of-the-art codecs (e.g., H.264/AVC [4], HEVC [5], including their scalable extensions), anchor motion fields at the (odd indexed) target frames of the temporal transform's prediction step; we refer to this as the *traditional anchoring* scheme. In this chapter, we flip the motion field anchoring and instead *hierarchically* anchor motion fields at the (even indexed) reference frames, which we refer to as *bidirectional hierarchical anchoring (BIHA)*. As we shall see later, this has some major advantages over the traditional motion field anchoring scheme. Figure 5.1 shows the traditional and the proposed BIHA schemes for $T = 3$ temporal transform levels.

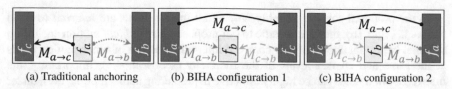

(a) Traditional anchoring   (b) BIHA configuration 1   (c) BIHA configuration 2

Fig. 5.2 Frame naming conventions used in the discussion. The target frame (in the middle) is predicted from its temporal left and right neighbour. **a** Shows the traditional anchoring; **b** and **c** show the two configurations that arise in the BIHA scheme, which depend on the index of the target frame

Note that the BIHA scheme requires the inversion of motion fields so that they can be used for temporal prediction of the target frames, as described in Sect. 4.2 of the previous chapter. To facilitate the discussion, we label the frames and motion fields involved in the bidirectional prediction process at any given temporal level t as shown in Fig. 5.2. In the BIHA scheme, there are two different arrangements of frames (Fig. 5.2b, c), depending on the index of the target frame. In Fig. 5.2b, one can identify the same motion field arrangement as the one used in BOA-TFI presented in Sect. 4.4. In fact, apart from the temporal hierarchical structure of the BIHA scheme, the main difference is that in a (scalable) coding scenario, the target frame f_b is known; furthermore, $M_{a \to b}$ and $M_{c \to b}$ can be estimated, so that they can account for acceleration. Besides this, the proposed highly scalable video compression scheme using BIHA motion employs the same operations as BOA-TFI to predict target frames.

In the following, we provide more insight into the aspects that are particular to a coding environment, and highlight the benefits of BIHA compared to the traditional anchoring of motion at target frames.

5.1.1 A Hierarchy of Scaled and Inferred Motion Fields

Both the traditional and proposed schemes involve a *full* motion field that is either independently coded, or differentially coded at a coarser temporal level. We use the terms *scaled* and *inferred* to refer to motion fields that serve as prediction references for motion field coding within the temporal hierarchy. In the proposed approach, the motion vectors of each motion field found at level t in the hierarchy are scaled by $\frac{1}{2}$ to form prediction references for a motion field at the next finer level $t + 1$. Scaled prediction references are also commonly used in the traditional motion anchoring approach, where the motion vectors of each reverse pointing motion field are scaled by -1 to serve as a prediction for the forward pointing motion field anchored at the *same* frame. The scaled motion fields are shown as dotted lines in Figs. 5.1 and 5.2. As prediction references, these scaled motion fields can be expected to be most efficient under constant (non-accelerated) motion.

In the proposed scheme, roughly *half* of all motion fields are *inferred* (dashed arrows in Fig. 5.1b); they are specific to our proposed *hierarchical* motion anchoring scheme, being obtained through composition and inversion of other motion fields at the same and coarser levels of the hierarchy (see Sect. 4.3). Importantly, the inferred motion fields can be highly effective in predicting actual motion, even under *accelerating* conditions, since they "follow" their scaled temporal sibling motion field. In the next section, we discuss the potential of motion field inference in the traditional anchoring scheme.

5.1.2   Potential for Motion Inference in the Traditional Anchoring Scheme

As a prediction tool, the inferred motion fields we compose in the proposed scheme are very appealing since they are very sparse – the prediction residual of inferred motion fields can be expected to be non-zero only in disoccluded regions. In this chapter, motion field inference is developed entirely for the proposed scheme. It is reasonable to ask whether the traditional scheme could potentially benefit from a similar approach.

As we shall see, the proposed approach makes it possible not only to infer motion fields, but also to deduce regions of disocclusion, where information in the target frame is not observable in one of the source frames. Without some form of explicit encoding, it is not clear how such information can be deduced in the traditional approach with target-frame anchored motion. It is important to note that it is not possible to have both scaled and inferred motion fields in the traditional scheme. Motion field scaling is a simple and effective mechanism to generate a prediction reference from another motion field that is anchored at the *same* frame. In the traditional approach, motion fields are anchored at the target frames, so that with the terminology of Fig. 5.2a, scaling can only be used to predict $M_{a \to b}$ from $M_{a \to c}$ or vice-versa.

Motion field inference could be used with the traditional anchoring scheme. In particular, with the terminology of Fig. 5.2a, $M_{a \to b}$ could be inferred from $M_{a \to c}$ and a coarser level motion field, or $M_{a \to c}$ could be inferred from $M_{a \to b}$ and a coarser level motion field, both of which are *alternatives* to motion scaling, but *not complementary*. Of course, to do this, the coarser level motion fields would also need to be encoded. The proposed approach has the benefit that scaling provides a robust motion prediction mechanism from coarse to fine levels, which is complemented by inference of the remaining finer level motion information, so that motion information need only be coded directly at the very coarsest level of the temporal hierarchy. Regardless of employing motion scaling or inference, the traditional anchoring requires one fully coded motion field (Fig. 5.1a) for each target frame; in contrast, there is only *one* fully coded motion field in the BIHA scheme (Fig. 5.1b).

5.1.3 Differential Coding of Motion Fields

In a coding scheme, the *scaled* and *inferred* motion fields serve as references $\hat{M}_{j\to i}$ for predictive coding of the actual motion field $M_{j\to i}$,

$$\Delta_{M_{j\to i}} = M_{j\to i} - \hat{M}_{j\to i}. \tag{5.1}$$

Clearly, the quality of these *scaled* and *inferred* prediction references has a large impact on the motion coding cost. For scaled motion fields, the *scaled motion residual* $\Delta_{M_{j\to i}}$ represents the *acceleration* between the three frames involved; *inferred motion residuals*, however, are expected to be non-zero only in regions that get disoccluded between frames f_a and f_c. The more temporal levels there are, the more efficient the scheme becomes, since the scaled and inferred residuals can be expected to become smaller at finer temporal levels. We use the term "highly" scalable to highlight the fact that the number of scalability levels does not need to be decided upon at the encoder. Furthermore, as presented in the last chapter, the proposed scheme is able to produce highly credible frames if all information at a given temporal level is quantized to zero, where it performs "intrinsic upsampling".

5.1.4 Geometrical Consistency with Quantized Motion Fields

As bits are discarded from a scalable bit-stream, small prediction residuals in $\Delta_{M_{j\to i}}$ will be quantized to zero, so that the motion obtained by the scaling and inference algorithms comes to dominate the visual properties of the reconstructed video. The proposed anchoring of motion fields at reference frames might appear counter-intuitive, because all motion fields have to be transferred to the target frames before motion-compensated prediction can be performed. One key insight of the proposed scheme is that because the *inferred* motion fields "follow" their *scaled* temporal sibling motion fields, the warped (inverse) motion fields, which are anchored at the target frame and used for the prediction of the target frame f_b, lead to geometrically consistent predictions, as shown in Fig. 5.3.

By contrast, in the case of the traditional anchoring, nothing guarantees that the forward and backward pointing motion fields point to the same geometrical location in the reference frames once the motion gets quantized, which leads to ghosting artefacts.

5.1.5 A Few Notes on Complexity

In addition to traditional MCTF as found in most video coders, the BIHA framework involves the following two main steps:

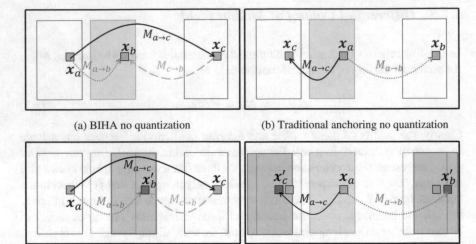

(a) BIHA no quantization   (b) Traditional anchoring no quantization

(c) BIHA quantized motion   (d) Traditional anchoring quantized motion

Fig. 5.3 Illustration of how quantization of motion fields affects the motion-compensated prediction process. **a–c** In the proposed BIHA scheme, the *inferred* motion field $M_{c \rightarrow b}$ "follows" whatever error there is in the *scaled* motion field $M_{a \rightarrow b}$. In contrast, **b, d** shows how in the traditional anchoring, the forward and backward pointing motion points to different geometrical locations, which leads to ghosting if the motion is quantized. This arises because $M_{a \rightarrow b}$ and $M_{a \rightarrow c}$ are not linked in the traditional anchoring

1. Transferring breakpoints from reference to target frames;
2. Transferring motion fields from one frame to another.

The proposed method is based on the fact that the underlying motion of a scene can be represented as being piecewise-smooth, with sharp transitions at motion field boundaries. For such motion fields, breakpoints can be expected to be sparse; since the complexity of the breakpoint warping procedure is linear in the number of breakpoints, the computational overhead of transferring breakpoints can be expected to be relatively low.

The core of both motion inversion and motion inference is the CAW procedure, which maps *motion* fields from one frame to another. For both the horizontal and vertical motion component, the complexity of this motion mapping is similar to the one of MCTF. For the experiments in this chapter, we used a fixed cell size of 1×1 pixels; hence, there are roughly twice as many triangular cells as there are pixels in the video. As mentioned earlier, we present a way of forming a mesh with adaptive triangle size in Sect. 7.2; here, we outline the main ideas of such a procedure and its consequences in a coding environment. Similar to the quadtree structure employed in modern hybrid video coders, the adaptive mesh implementation uses a *hierarchical cell* structure. Whenever a cell contains a nonzero motion wavelet coefficient, it is split up into 4 smaller cells, until the (sub)cell is smooth. If we let the maximum cell size be 32×32, then in the worst case, one nonzero wavelet coefficient creates

5×4 cells. Normally, such nonzero wavelet coefficients are grouped together around moving object boundaries, and multiple coefficients cause the same cell to be further partitioned. In the case of a truly isolated motion coefficient, the associated coding cost can be expected to be high. Assuming 10 bits to code this motion coefficient, the maximum number of partitioned cells per coded motion bit would be $\frac{5 \times 4}{10} = 2$. The number of cells N_c to be expected is thus linked to the motion bit-rate r_m; in practice, we expect $N_c \ll r_m \times 2$. Even this conservative bound suggests typical cell sizes to involve many pixels at reasonable motion bit-rates.

5.2 Motion-Compensated Temporal Filtering

For the coding part, we assume the use of a 5/3 *temporal* wavelet decomposition, based on motion-compensated lifting steps. At any given temporal level in this transform, the odd indexed frames are predicted using the preceding and succeeding even indexed frames, while even indexed frames are updated using the prediction residuals from their even indexed temporal neighbours. The even indexed frames are interpreted as a low-pass temporal subband and the procedure is recursively applied to this subband for a total T levels. This transform is common in the literature [6].

Both original and inverted motion fields are used together with *discovered* information about disoccluded regions, to drive a motion compensated temporal lifting transform of the video texture information. The temporal transform employed in this work is composed of two parts:

1. Bidirectional *prediction* of the target frame from its temporal neighbours;
2. A temporal *update* step, which feeds some of the motion compensated residual from the prediction step back to the reference frames.

The *update* step is specific to a compression scheme, and helps reduce temporal aliasing in case finer temporal levels are discarded from the bit-stream; it also has fundamental benefits in reducing the impact of quantization noise in the temporal subbands on reconstructed video quality [7].

In the proposed framework, the prediction step uses inverted motion fields, while the update step uses the original (encoded) motion fields. This differs markedly from traditional approaches, where bidirectional prediction is performed using original motion fields. However, the proposed approach has the advantage that disocclusion information, *discovered* during the inversion process, can inform the prediction process, as discussed in Sect. 4.2.1. In the following, we use the notation for disocclusion masks introduced in (4.5).

In the BIHA scheme, we compute two such disocclusion masks: one during the inversion of $M_{a \to b}$, and the other one while inverting $M_{c \to b}$. These masks are denoted $\hat{S}_{b \to a}$ and $\hat{S}_{b \to c}$, respectively.

5.2.1 Prediction Step

The inverted motion fields $\hat{M}_{b \to a}$ and $\hat{M}_{b \to c}$ are used to warp the texture from the reference frames f_a and f_c, respectively, to the target frame f_b; we use $f_{i \to j}$ to denote the frame obtained by warping the texture of frame f_i to frame f_j. Then, each (integer) location $\mathbf{m}$ in frame f_b is bidirectionally predicted as

$$\hat{f}_b[\mathbf{m}] = \begin{cases} \dfrac{\hat{S}_{b \to a}[\mathbf{m}] f_{a \to b}[\mathbf{m}] + \hat{S}_{b \to c}[\mathbf{m}] f_{c \to b}[\mathbf{m}]}{\kappa[\mathbf{m}]} & \kappa[\mathbf{m}] > 0 \\ 0.5\big(f_{a \to b}[\mathbf{m}] + f_{c \to b}[\mathbf{m}]\big) & \kappa[\mathbf{m}] = 0 \end{cases}, \qquad (5.2)$$

where $\kappa[\mathbf{m}] = \hat{S}_{b \to a}[\mathbf{m}] + \hat{S}_{b \to c}[\mathbf{m}]$. The prediction residual Δf_b that needs to be coded is

$$\Delta f_b = f_b - \hat{f}_b, \qquad (5.3)$$

which is also the *high-pass temporal subband*.

5.2.2 Update Step

Motion fields are invertible everywhere except in disoccluded regions. We use the disocclusion masks $\hat{S}_{b \to j}$, where $j \in \{a, c\}$ are the previous and future reference frames, to disable the update step in disoccluded regions. Defining the *update weight* β as

$$\beta = \begin{cases} 0.25 & \kappa(\boldsymbol{x}) = 2 \\ 0.5 & \kappa(\boldsymbol{x}) < 2 \end{cases}, \qquad (5.4)$$

the updated frame becomes

$$f_j^{\text{updated}}(\mathbf{m} + \hat{M}_{b \to j}[\mathbf{m}]) = f_j(\mathbf{m} + \hat{M}_{b \to j}[\mathbf{m}]) + \beta \hat{S}_{b \to j}[\mathbf{m}] \Delta f_{b_{b \to j}}[\mathbf{m}]. \qquad (5.5)$$

We remind the reader that we use $(\cdot)$ and $[\cdot]$ to differentiate between a *continuous* and an *integer* location, respectively. While the update step is performed for all *integer* locations $\mathbf{m}$ in the prediction residual Δf_b of the target frame, the update is fed back using motion $\hat{M}_{b \to j}$, which normally does not fall onto integer locations in the reference frame f_j. We use a bilinear weighting to proportionally assign the prediction residual to its closest neighbours on the integer grid of the respective reference frame.

Equations (5.4) and (5.5) imply that a quarter of the prediction residual is fed back to the two reference frames in regions that are visible from both references. If a location is only visible from one side, the temporal transform is effectively reduced from the 5/3 temporal wavelet transform to a 2-tap Haar transform. If a region is disoccluded in both frames, (5.5) eliminates the update step; in such regions the prediction step in (5.2) averages two spatially smooth predictors.

5.3 Scalable Rate Allocation

In order to quantitatively evaluate the proposed anchoring and compare it to the traditional anchoring of motion fields, we have to code the texture, motion, and breakpoint data. For this evaluation to be meaningful, we need to balance the error contributions of the different data types, which requires an understanding on how errors propagate. In the following, we present an analytical model that allows us to understand how quantization errors in the motion field subbands impact distortion in the final reconstructed video sequence. We then consider the impact of errors in the texture and breakpoint information in Sect. 5.3.2.

5.3.1 Motion Error

We first investigate how a quantization error of δ propagates across frames for the different types of motion fields. Let $f^{(t)}$ denote the frames produced after $T - t$ levels of temporal synthesis. For any level $t \in [0, T - 1]$ of the temporal transform, motion fields $\{M_{i \to j}^{(t)}\}$ are used to synthesize frames $f^{(t)}$ from frames $f^{(t+1)}$ together with high-pass temporal subband frames $d^{(t+1)}$.

For any level t, the reconstructed frames at the finest temporal level can always be expressed as

$$f^{(0)} = S_L^{(t+1)}(f^{(t+1)}) + \sum_{p=1}^{t+1} S_H^{(p)}(d^{(p)}), \tag{5.6}$$

where $S_L^p(\cdot)$ and $S_H^p(\cdot)$ denote low- and high-pass temporal synthesis operators associated with the information injected at level p in the transform.

For convenience of analysis, consider for the moment that there is a constant displacement error δ in $M_{i \to j}^{(t)}$. Our goal is to understand the impact of this error on the total squared error distortion in the reconstructed video. This reconstructed video distortion arises from geometric distortions (i.e., spatial shifts) in each of the synthesized texture contributions found in (5.6). In particular, the synthesis operators $S_L^{(p)}$ and $S_H^{(p)}$, $p \leq t + 1$ each depend on $M_{i \to j}^{(t)}$ and hence on the error δ, so that the total error experienced from all frames at the finest temporal level becomes

$$\Delta f^{(0)} = D_L^{(t+1)}(f^{(t+1)}, \delta) + \sum_{p=1}^{t+1} D_H^{(p)}(d^{(p)}, \delta), \tag{5.7}$$

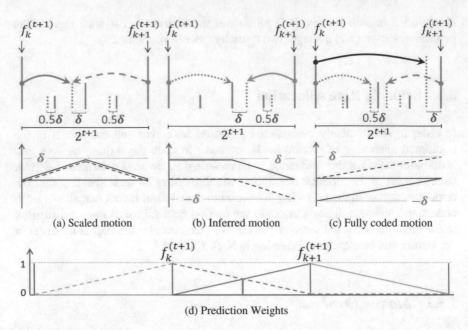

(a) Scaled motion   (b) Inferred motion   (c) Fully coded motion

(d) Prediction Weights

Fig. 5.4 Illustration how errors in motion fields spread in the temporal transform for **a** scaled, **b** inferred, and **c** fully coded motion fields. The solid red line shows how the texture data from the right reference frame is affected by introducing an error δ into a motion field; the dashed red line shows the same for the left reference frame. **d** Shows the prediction weight of the two reference frames

While it is possible to analyze each of these contributions separately, it turns out that the major contribution to $\Delta f^{(0)}$ arises from the first term $D_L^{(t+1)}(f^{(t+1)}, \delta)$, corresponding to distortion arising for the low-pass synthesis operation. In the following analysis, we therefore assume that all detail bands $d^{(p)}$ are zero. In order to simplify the ensuing analysis, we consider only the case where the original motion is 0, so that the error δ becomes the distorted motion field. The resulting analysis remains valid for any translational original motion field, and is an excellent approximation for more general original motion fields.

Figure 5.4 shows the effect of an error δ in any level t scaled, inferred or full motion fields found between frames $f_k^{(t+1)}$ and $f_{k+1}^{(t+1)}$. The second row in the figure shows how the texture contributions from these left (dashed line) and right (solid line) reference frames become shifted by the time they reach the finest temporal level $f^{(0)}$. Figure 5.4d shows the relative contribution from each of $f_k^{(t+1)}$ and $f_{k+1}^{(t+1)}$ to each frame of the final synthesized video sequence.

In the following, we provide an asymptotic analysis of the impact of δ in the limit as t becomes very large so that the synthesized video can be treated as continuous in time. As revealed by Fig. 5.4, the motion errors in question result in shifted contributions that can produce reconstructed video errors only at frame times $(k+\tau)2^{t+1}$, where $\tau \in [0, 1]$. We can express these error frames in terms of their contributions from the left and right reference frames as $\Delta f_\tau^{(0)} = \Delta f_{\text{left},\tau}^{(0)} + \Delta f_{\text{right},\tau}^{(0)}$.

5.3.1.1 Scaled Motion Fields

The scaled motion field $M_{2k \to 2k+1}^{(t)}$, shown as a solid blue arrow in Fig. 5.4a, has motion vectors $\mathbf{u}_{2k \to 2k+1}(\mathbf{x})^{(t)}$. Introducing an error of δ to these motion vectors yields an error contribution $\Delta f_{\text{left},\tau}$ that can be expressed in the Fourier domain as

$$\Delta \hat{f}_{\text{left},\tau}(\boldsymbol{\omega}) = \hat{f}(\boldsymbol{\omega})\big(1 - e^{-j\boldsymbol{\omega}^t \delta 2\tau}\big)(1 - \tau), \tag{5.8}$$

over the interval $\tau \in [0, 0.5]$. Here, $\hat{f}(\boldsymbol{\omega})$ is the Fourier transform of $f_k^{(t+1)}$, which is identical to $f_{k+1}^{(t+1)}$ under our zero original motion assumption. As shown in Fig. 5.4a, the same shifts arise in the contribution from $f_{k+1}^{(t+1)}$ due to the motion inference process. Accordingly,

$$\Delta \hat{f}_{\text{right},\tau}(\boldsymbol{\omega}) = \hat{f}(\boldsymbol{\omega})\big(1 - e^{-j\boldsymbol{\omega}^t \delta 2\tau}\big)\tau. \tag{5.9}$$

Thus, for $\tau \in [0, 0.5]$, the total error at frame $f_\tau^{(0)}$ can be expressed as

$$\Delta \hat{f}_\tau(\boldsymbol{\omega}) = \hat{f}(\boldsymbol{\omega})\big(1 - e^{-j\boldsymbol{\omega}^t \delta 2\tau}\big) \approx \hat{f}(\boldsymbol{\omega}) 2\tau j \boldsymbol{\omega}^t \delta, \tag{5.10}$$

where we have used a first order Taylor series approximation for the complex exponential.

Evidently, the errors $\Delta \hat{f}_\tau(\boldsymbol{\omega})$ that arise when $\tau \in [0.5, 1]$ are just a mirror image of those that arise when $\tau \in [0, 0.5]$. Using Parseval's theorem, the total energy of the prediction error $|e_{scal}^\infty|^2$ can then be expressed as

$$|e_{\text{scal}}^\infty|^2 = 2^{t+1} \cdot 2 \int_0^{0.5} \frac{1}{(2\pi)^2} \int_{-\pi}^{\pi} \int_{-\pi}^{\pi} |\Delta \hat{f}_\tau(\boldsymbol{\omega})|^2 \mathrm{d}\tau \mathrm{d}\boldsymbol{\omega} \tag{5.11}$$

where the factor of 2^{t+1} arises from the observation that our interval $\tau \in [0, 1]$ corresponds to 2^{t+1} reconstructed video frames. This error energy (distortion) can be approximated by

$$|e_{\text{scal}}^\infty|^2 \approx 2^{t+4} \underbrace{\int_0^{0.5} \tau^2 \mathrm{d}\tau |\delta|^2 \frac{1}{(2\pi)^2} \int_{-\pi}^{\pi} \int_{-\pi}^{\pi} |\hat{f}(\boldsymbol{\omega})|^2 |\boldsymbol{\omega}|^2 \cos^2(\Theta) \mathrm{d}\boldsymbol{\omega}}_{= \frac{1}{2} E[|\nabla f|^2] A \text{ (assuming isotropic power spectrum)}}$$

$$= \frac{2^t}{3} E[|\nabla f|^2] A |\delta|^2, \tag{5.12}$$

where $E[|\nabla f|^2]$ is the average gradient power, and A is the area of the frame. Let $D_{\mathbf{u}^{(t)}}^{\text{scal}}$ denote the total amount of distortion in scaled motion fields at temporal level t. The resulting total amount of distortion in the reconstructed video can then be expressed as

$$D_{\mathbf{u}^{(t)} \to f_0}^{\text{scal}} = D_{\mathbf{u}^{(t)}}^{\text{scal}} \frac{2^t}{3} E[|\nabla f|^2] = D_{\mathbf{u}^{(t)}}^{\text{scal}} \cdot \alpha_{\text{scal}}^{(t)} \cdot E[|\nabla f|^2]. \qquad (5.13)$$

While the model we present assumes a constant error δ, we note that it provides a good approximation also for gradually changing errors. We do not specifically consider high frequency motion errors, but note that the sparse motion representation required for good coding efficiency inevitably leads to motion fields (and hence motion quantization errors) that are smooth except in the vicinity of breakpoints.

5.3.1.2 Inferred Motion Fields

The inferred motion field $M_{2k+2 \to 2k+1}^{(t)}$, shown as a solid orange arrow in Fig. 5.4b, has motion vectors $\mathbf{u}_{2k+2 \to 2k+1}(\boldsymbol{x})^{(t)}$. Errors in these motion vectors do not affect the scaled sibling motion field $M_{2k \to 2k+1}^{(t)}$. This can be seen in Fig. 5.4b, where the dashed red line, indicating the shift in the texture of the left reference frame, is zero over the first half interval $\tau \in [0, 0.5]$. Over this half interval, $\hat{f}_\tau$ can be expressed in the Fourier domain as

$$\Delta \hat{f}_\tau(\omega) = \hat{f}(\omega)\big(1 - (1 - \tau) - \tau e^{-j\omega^t \delta}\big) \approx \hat{f}(\omega)\tau j\omega^t \delta. \qquad (5.14)$$

For $\tau \in [0.5, 1]$, the expression is

$$\Delta \hat{f}_\tau(\omega) = \hat{f}(\omega)\big(1 - (1 - \tau)e^{j\omega^t \delta(2\tau-1)} - \tau e^{j\omega^t \delta(2\tau-2)}\big)$$
$$\approx \hat{f}(\omega)(\tau - 1)j\omega^t \delta. \qquad (5.15)$$

Again, using Parseval's Theorem, the sum of squared errors can be written as

$$|e_{\text{inf}}^\infty|^2 \approx \frac{1}{2} E[|\nabla f|^2] A 2^{t+1} |\delta|^2 \Big(\int_0^{0.5} \tau^2 d\tau + \int_{0.5}^1 (\tau - 1)^2 d\tau \Big) \qquad (5.16)$$
$$= \frac{2^t}{12} E[|\nabla f|^2] A |\delta|^2.$$

Let $D_{\mathbf{u}^{(t)}}^{\text{inf}}$ denote the total amount of distortion in inferred motion fields at temporal level t. The resulting total amount of distortion in the reconstructed video can then be expressed as

$$D_{\mathbf{u}^{(t)} \to f_0}^{\text{inf}} = D_{\mathbf{u}^{(t)}}^{\text{inf}} \frac{2^t}{12} E[|\nabla f|^2] = D_{\mathbf{u}^{(t)}}^{\text{inf}} \cdot \alpha_{\text{inf}}^{(t)} \cdot E[|\nabla f|^2]. \qquad (5.17)$$

5.3.1.3 Full Motion Fields

At the coarsest level of the temporal hierarchy, where $t = T - 1$, the proposed method involves one full motion field $M_{k \to k+1}^{(T)}$, with motion vectors $\mathbf{u}_{k \to k+1}(\mathbf{x})^{(T)}$, as shown in Fig. 5.4c. Full motion fields are never used to directly predict their target frame, but all lower level motion fields depend upon them. An error of δ in a full motion field leads to error contributions

$$\begin{aligned}
\Delta \hat{f}_{\text{left},\tau}(\boldsymbol{\omega}) &= (1 - \tau) \hat{f}(\boldsymbol{\omega}) \left(1 - e^{-j\omega^t \delta \tau}\right) \\
\Delta \hat{f}_{\text{right},\tau}(\boldsymbol{\omega}) &= \tau \hat{f}(\boldsymbol{\omega}) \left(1 - e^{-j\omega^t \delta(\tau - 1)}\right)
\end{aligned} \qquad (5.18)$$

Each term alone is a stretched version of the corresponding term that we studied in connection with scaled motion fields. Indeed, if we consider only the left or right error contribution in isolation, the total squared error associated with such a contribution turns out to be the same for both full motion field errors and scaled motion field errors at level $t = T - 1$. However, the left and right error contributions from an error in the full motion field approximately cancel each other out. This is because geometric shifts in the left contribution are matched by opposing shifts in the right contribution, as seen in the second row of Fig. 5.4c. It follows that full motion field errors produce significantly smaller levels of reconstructed video distortion than errors in the scaled motion fields. However, the divergent shifts induced in the left and right reference frame contributions to $f^{(0)}$ yield substantial levels of "ghosting." By contrast, distortions introduced by errors in scaled motion fields are free from such visually disturbing ghosting artefacts. It would be beneficial to adopt a distortion metric which could specifically account for the objectionable nature of ghosting artefacts; however, the development of such a metric would require subjective evaluations that lie beyond the scope of this thesis. As a compromise, therefore, we choose to assign the same weighting factor to both scaled and full motion fields, i.e., $\alpha_{\text{full}}^{(T)} = \alpha_{\text{scal}}^{(T-1)}$, leaving us with the model

$$D_{\mathbf{u}^{(T)} \to f_0}^{\text{full}} = D_{\mathbf{u}^{(T)}}^{\text{full}} \cdot \alpha_{\text{full}}^{(T)} \cdot E[|\nabla f|^2]. \qquad (5.19)$$

5.3.1.4 Distortion Scaling Factors for the Discrete Case

The asymptotic analysis above unveils the coupling between motion errors and reconstructed video errors for the proposed motion representation. For small t, where very few video frames lie in the interval $[k \cdot 2^{t+1}, (k + 1) \cdot 2^{t+1}]$, this continuous analysis is only approximately valid. Actual coupling factors are shown in Table 5.1 for $T = 4$ levels of temporal decomposition. Squared quantization errors in individual

Table 5.1 Squared errors for different temporal texture and motion subbands for a total of $T = 4$ temporal decompositions

t	$\alpha_{\text{text}}^{(t)}$	$\alpha_{\text{scal}}^{(t)}$	$\alpha_{\text{inf}}^{(t)}$	$\alpha_{\text{full}}^{(t)}$
0	0.719	0.5	0.125	–
1	0.922	0.75	0.1875	–
2	1.586	1.375	0.34375	–
3	3.043	2.6875	0.671875	–
4	–	–	–	2.6875

subbands b of the breakpoint-adaptive spatial wavelet transform that is used to compress motion fields of type "mtyp" are scaled by the overall weighting factor

$$w_{\text{mtyp}}^{(t,b)} = \alpha_{\text{mtyp}}^{(t)} \cdot E[|\nabla f|^2] \cdot G_{\text{mdwt}}^{(b)} \tag{5.20}$$

in order to discover their impact on reconstructed video distortion; here $G_{\text{mdwt}}^{(b)}$ is the squared Euclidean norm of the spatial DWT synthesis basis functions associated with motion subband b.

5.3.2 Rate Allocation with Breakpoint and Texture Errors

Breakpoints and motion are tightly linked, and hence the analysis for errors introduced by quantizing breakpoints is similar to the one presented above. More details can be found in [8]. As we will see in Sect. 5.4.1, this approximation leads to a very good performance.

The temporal texture subbands produced by our motion adaptive temporal transform are also subjected to a spatial DWT, whose subband samples are subject to quantization errors. The impact of squared subband quantization errors on distortion in the reconstructed video sequence can be modelled using a separate set of weighting factors

$$w_{\text{text}}^{(t,b)} = \alpha_{\text{text}}^{(t)} G_{\text{tdwt}}^{(b)} \tag{5.21}$$

where $G_{\text{tdwt}}^{(b)}$ is the squared Euclidean norm of the spatial DWT synthesis functions associated with texture subband b, and $\alpha_{\text{text}}^{(t)}$ is the squared Euclidean norm of the temporal synthesis basis functions associated with temporal subbands at level t.

Together, these weights are used to drive a rate-distortion optimized rate allocation algorithm. In practice, the rate allocation is performed using *the post compression rate-distortion (PCRD)* strategy of JPEG2000's EBCOT algorithm [9]. That is, each of the individual subbands and breakpoint vertex bands are subjected to embedded block-based coding, collecting distortion-length slopes for each coding pass, after which the block coding passes are arranged into a global set of quality layers based on the distortion-length slopes, weighted using the factors found above. The resulting scalable video bit-stream can be reconstructed at any of the rate-distortion optimal operating points obtained by discarding quality layers from the overall representation.

5.4 Hierarchical, Spatio-Temporal Induction of Discontinuity Information

In Sect. 4.1, we presented a temporal breakpoint induction method, which is used to transfer breaks from one frame to another. In the case of *temporal frame interpolation (TFI)*, by definition, there is no information present at the target frame; in particular, there is no motion discontinuity information. In this section, we augment the temporal breakpoint induction to a *hierarchical spatio-temporal breakpoint induction (HST-BPI)* method to account for coded spatial breaks.

A natural outcome of the proposed hierarchical coding framework is that at the decoder, the precision of texture, motion, and breakpoint data is higher at coarser temporal levels t (see Sect. 5.3); the same is true for spatial resolutions η. One can therefore expect that at lower bit-rates, few if any *vertices* will appear at the finer spatio-temporal resolution levels. Since both spatial and temporal induction processes are of interest, we must be able to resolve conflicts between the induced breakpoints that may arise. In this work, we perform induction in a particular sequence, in which breakpoints are induced to all spatial levels of the frames at a particular level in the temporal hierarchy, before moving to the next finer temporal level; this is visualized in Fig. 5.5.

HST-BPI is hierarchical in that it goes from coarse to fine spatial levels. At each spatial level η, all cells are traversed; for cells that contain two breakpoints on perimeter arcs, "discontinuity" line segments are formed. For each such line segment, the following three steps are performed:

1. *Breakpoint compatibility check (BCC)* to find *compatible* (i.e., foreground) motion to assign to discontinuity line segments;
2. Warping of temporally compatible line segments (Temporal induction);
3. Upsampling of all breakpoints to the next finer spatial resolution (Spatial induction).

The breakpoint compatibility check in step 1 of the proposed procedure is almost identical to the first step of the temporal induction procedure presented in Sect. 4.1. The only difference is that it is performed from coarse to fine spatial levels, instead of just at the finest spatial resolution.

In a coding environment, (partial) breakpoint information that was estimated at the target frame f_b can be present; this is particularly true at higher bit-rates. Since such *spatial* breaks were estimated on an actual motion field, we want to give them higher priority than the temporally induced ones, which we refer to as *temporal* breaks. In particular, for each intersection of the warped line segment l_b with an

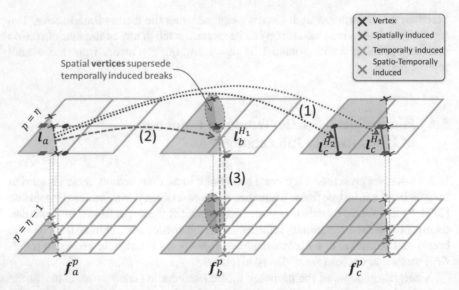

Fig. 5.5 Hierarchical, spatio-temporal induction of breakpoints (HST-BPI). Going from coarse to fine spatial resolution, the proposed method consists of three steps at each resolution level η: (1) Assessment of temporal compatibility of line segments induced by breakpoints between two coarse-level frames f_a and f_c; (2) Warping of compatible line segments to f_b; (3) Spatial induction of all breakpoints to the next finer spatial resolution $\eta - 1$. For better visualization, root arcs are not shown in this figure

arc in f_b, we check whether the temporally induced break falls into a *spatial break occupied* (SBO) cell. A cell is considered SBO if it contains at least one *spatial break*. If the cell is empty or only contains (spatio-)temporally induced breaks (i.e., *not* SBO), the temporal break is registered at the position of the intersection of the line segment with the arc, replacing any existing breakpoint on that arc. Note that spatial breaks can never be replaced by this scheme, because any arc containing a spatial break necessarily belongs to an SBO cell. In the third step, all breakpoints are transferred to the next finer spatial level, where spatial induction is performed to *induce* breakpoints to the root arcs.

5.4.1 Evaluation of HST-BPI

This section evaluates the HST-BPI in a coding scenario. As mentioned in Sect. 5.4, in a compression scenario, the aim of HST-BPI is to *improve* existing (spatial) breakpoint fields. At very high bit-rates, high quality spatial breakpoint fields are anchored at the target frames, and temporal induction should ideally not change anything. At medium to low bit-rates, only few or no spatial breakpoints might be decoded at fine temporal levels; in this case, the scheme completely relies on temporally induced breakpoints.

Fig. 5.6 Evaluation of the proposed HST-BPI method on three synthetic test sequences. The Y-PSNR of the reconstructed frame f_1 is obtained by decoding different levels of spatial breakpoints at f_1: The dashed curve is obtained by decoding all spatial breaks at f_1; the dotted curve shows the results if *no* spatial breakpoints are decoded at f_1, and hence relying on temporal breakpoint induction; the solid curve shows the R-D performance if breakpoints are scalably decoded with respect to the quality of the motion field

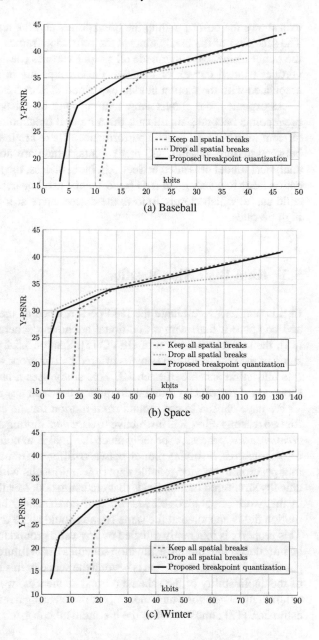

(a) Baseball

(b) Space

(c) Winter

In order to evaluate the proposed HST-BPI with residual coding, we perform one level of temporal decomposition followed by 5 levels of spatial breakpoint-adaptive DWT of the texture and motion data; the wavelet coefficients are then coded using EBCOT [10]. Breakpoints are estimated based on the motion fields, and coded as explained in [8]. The scalable bit-stream is then decoded at various quality levels. Figure 5.6 shows the PSNR of the reconstructed target frame f_1, and the horizontal axis shows the cost of coding the breakpoints and the texture residual at frame f_1. The precision of the breakpoints at the reference frames f_0 and f_2 is kept high; for the target frame, we either code all spatial vertices (dashed curve), code no spatial vertices (dotted curve), or quantize the breakpoints in accordance to the quality associated with the motion field as explained in Sect. 5.3.2 (solid curve).

As expected, the graphs show that at lower bit-rates, where the cost of coding breakpoints becomes significant, the temporal breakpoint induction leads to a significant improvement in R-D performance. There are many complex dependencies between texture, motion, and breakpoints, which are not all accounted for by the analytical model presented in Sect. 5.3. Nonetheless, the resulting scalable rate allocation leads to a very good R-D performance at all bit-rates; this is evidenced by the solid curve, which closely follows the dotted curve at low, and the dashed curve at high bit-rates.

5.5   Rate-Distortion Results

In this section, we evaluate the R-D performance of the proposed BIHA scheme, and compare it both with a traditional anchoring scheme, as well as with SHVC [11], the scalable extension of HEVC [5]. Section 5.5.2 provides a comprehensive study of the rate-distortion benefits offered by our proposed approach, including the rate allocation scheme of Sect. 5.3. This study uses a large collection of synthetic sequences,[2] for which ground truth motion between *any pair of frames* is available.

We have chosen to use ground truth motion for the comprehensive comparison for two reasons. First, it is instructive to *decouple* motion *estimation* from the motion *compensation* process, especially since the goal is to compare quite different transform structures, involving motion between different frame pairs. Also, we are not yet in a position to provide reliable experimental results with a complete set of hierarchically structured motion fields that are estimated. Ideally, motion fields employed in this work should exhibit *hierarchical consistency*, including the property that motion fields anchored at the same frame should share the same set of breakpoints. This property is inherently satisfied by any valid ground truth motion field and could be introduced into motion estimation schemes in the future. This is an interesting and parallel stream of research that is beyond the scope of this thesis. To provide evidence of the applicability of the scheme on real sequences, we apply the method on two natural sequences, for which motion, estimated using a readily available optical flow estimator, [12] almost satisfies the hierarchical consistency.

---

[2] Available on: http://ivmp.unsw.edu.au/~dominicr/biha_scheme.html.

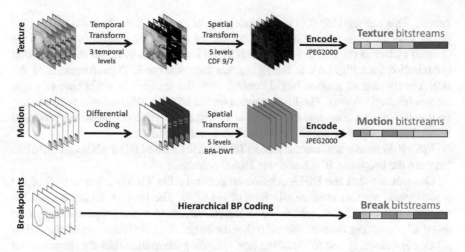

Fig. 5.7 Experimental setup used for the evaluation of BIHA in a highly scalable video compression system

5.5.1 Experimental Setup

Figure 5.7 gives an overview of the experimental setup. For both the proposed BIHA and the traditional anchoring (TRAD) scheme, the sample sequences are compressed using $T = 3$ levels of temporal decomposition, resulting in 8 frames per GOP. The temporal subband frame textures are then subjected to $D = 5$ levels of spatial DWT, and the differentially coded motion fields are subjected to $D = 5$ levels of spatial BPA-DWT. The quantized wavelet coefficients are coded using EBCOT [10]. Breakpoints are coded using the method described in [8], and quantized based on the quality of the motion fields they are coding; intuitively, the more quantized the motion fields, the less breakpoints there are. We remind the reader that all elements of the coded representation are highly scalable. The results are obtained by weighting motion and texture subbands according to (5.20) and (5.21), respectively; appropriate weights were also computed for the traditional anchoring scheme.

For SHVC (SHM version 7.0 (HM-15.0)), we used the main profile of the random access encoder (hierarchical B-frames), and created a base layer at QP 38 at half the native resolution of the input sequence, and 5 enhancement layers at full resolution at QPs {23, 26, 30, 34, 38}.

5.5.2 R-D Comparisons with Traditional Anchoring Scheme

We evaluate the rate distortion performance of the proposed BIHA scheme, and compare it with the traditional way (TRAD) of anchoring motion fields at target frames. This comparison provides a good way of analyzing the benefits of the proposed scheme. Figure 5.8 shows R-D curves for the three synthetic sequences we showed earlier in the evaluation of the HST-BPI scheme to evaluate the breakpoint quantization (see Fig. 5.6). In the figure, we compare the R-D performance of the BIHA anchoring of motion (solid curves) with the traditional anchoring at target frames (dashed curves). The filled regions on the left show the bit-rate used for just coding the motion data; one can see how at medium to high bit-rates, the proposed hierarchical anchoring enables more efficient motion coding.

More R-D results are summarized in Table 5.2 in terms of BD-PSNR and BD-Rate between the proposed BIHA and the TRAD scheme.

One can see that the BIHA scheme outperforms the TRAD scheme in 8 of the 9 sequences, with an average BD-Rate of −14.8%. The bit-rate saving on just the motion fields is −13.2%, which shows the effectiveness of the proposed method in terms of predicting motion. We note that the better R-D performance of the BIHA scheme is not solely due to the lower cost of coding the motion, but also because our scheme is able to produce *geometrically consistent* predictions, even using quantized motion (see Fig. 5.10b, d, f for examples); on average, the BD-rate on just the texture data is −17.5%. One can see that the BIHA scheme performs worse in the "Flowers" sequence. This sequence, containing significant rotation of the background, is particularly affected by a shortcoming of the proposed background motion extrapolation technique in disoccluded regions. The problem is that we currently extrapolate background motion for each triangle individually, which creates artificial boundaries in the disoccluded region, and these are expensive to code. In future work, we plan to address this issue by performing the background extrapolation on connected disoccluded regions rather than individual triangles, which will avoid such artificial boundaries. Nonetheless, even without such improvement, the proposed scheme clearly outperforms the TRAD method on balance.

5.5.3 Importance of Motion Discontinuities

Many of the benefits of the proposed BIHA framework come from the use of *motion discontinuities*. A natural question to ask is what happens if they are removed; this question is particularly relevant in the presence of motion blur. In [13], we proposed a way of including motion blur synthesis into the BIHA scheme. For conciseness, we will not delve into the details of this work. For the motion blur synthesis work, we generated sequences that are heavily affected by motion blur. In this section, we use some of the sequences generated for that work to show the importance of motion discontinuities even on sequences that do not contain sharp transitions in the texture data.[3]

---

[3]Wulff and Black [14] show that piecewise-smooth motion fields with sharp discontinuities can be estimated on sequences that are heavily affected by motion blur.

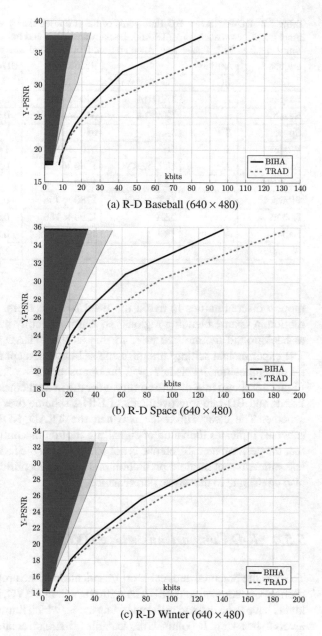

Fig. 5.8 Average per-frame bit-rate against PSNR for the same synthetic sequences as the one we used in Fig. 5.6, obtained using $T = 3$ temporal decomposition levels (GOP size $= 8$). The filled regions on the left show the average bit-rate spent on coding motion field data for the BIHA (dark gray) and TRAD (light gray) schemes

(a) R-D Baseball (640×480)

(b) R-D Space (640×480)

(c) R-D Winter (640×480)

As a first step, we removed discontinuities from the BIHA framework, but the results were too poor to be reported. Instead, it is instructive to consider the *traditional* motion anchoring approach, where motion is anchored at the target frames. In that case, motion fields do not have to be inverted, and hence one can consider removing

Table 5.2 BD-PSNR and BD-Rate gains of the proposed BIHA scheme compared to the traditional target-based anchoring (TRAD). Background (BG) and foreground (FG) motion activity is summarized as **S**till, **T**ranslation, **A**cceleration, **R**otation, and **Z**oom

Sequence	Activity		Resolution	BD-PSNR (dB)	BD-Rate (%)
	BG	FG			
Baseball	A	1A, 1RA	640 × 480	3.35	−34.84
Beach	S	1ZA, 2A	640 × 480	0.97	−12.46
Space	S	3A	640 × 480	1.86	−24.17
Winter	S	5A	640 × 480	1.41	−14.55
Autumn	A	2RT, 1ZRT	1280 × 736	0.78	−9.82
Butterfly	T	1ZA	1280 × 736	1.58	−18.41
Flowers	R	1RT	1280 × 736	−0.34	5.27
Robots	A	2ZA	1280 × 736	0.63	−12.71
Balls	A	3RA	1920 × 1088	0.82	−11.72
Average	-	-	-	1.23	−14.82

motion discontinuities to avoid the need to code them. Furthermore, we blurred motion in texture blending regions[4] so that it smoothly transitions from foreground to background motion; we refer to this as the "TRAD_NOBREAKS" approach. The experimental settings are the same as before, except that the number of temporal decomposition levels is $T = 2$ (as opposed to $T = 3$ in the other experiments). Figure 5.9 shows rate-distortion curves for the two schemes, and Table 5.3 shows BD-PSNR and BD-rate improvements of BIHA scheme over the TRAD_NOBREAKS scheme. The clear difference between the TRAD_NOBREAKS and the BIHA schemes indicates the value of signalling motion discontinuities. It is worth noting that the performance difference is not solely due to the lack of motion blur handling; less efficient motion field prediction and lack of identification of occluded regions also contribute to the worse performance.

5.5.4   R-D Comparisons with SHVC

To show the potential and real-world application of the proposed scheme, we provide preliminary comparisons of BIHA with SHVC. In SHVC, the base and enhancement layers have to be defined at the encoding stage, which limits the number of scalability levels to just a few. In contrast, the scalable bit-stream created in the BIHA and TRAD schemes can be truncated at any bit-rate, enabling a *highly* scalable framework. For this reason, BD-Rate/PSNR comparisons between these entirely different schemes are not very meaningful. Instead, we provide R-D curves for three synthetic sequences

---

[4]These are regions around moving objects, where the foreground and background texture information is mixed together.

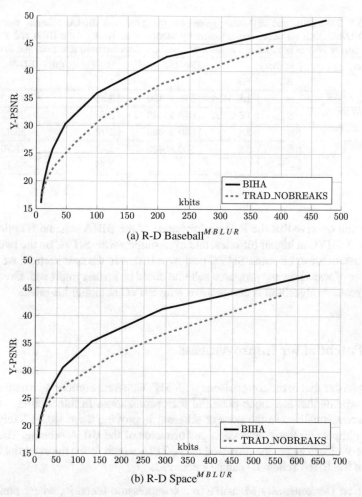

(a) R-D Baseball^{MBLUR}

(b) R-D Space^{MBLUR}

Fig. 5.9 Average per-frame bit-rate against PSNR for two synthetic scenes affected by motion blur (MBLUR), obtained using $T = 2$ temporal decomposition levels. We compare the BIHA scheme with the traditional anchoring of motion fields at target frames, as well as motion fields anchored at reference frames with smooth motion in texture blending regions (TRAD_NOBREAKS)

in Fig. 5.10, as well as two for natural test sequences in Fig. 5.11; in both figures, the limited scalability of SHVC is indicated by the staircase curve in the figure.

The distortion is expressed in terms of average Y-PSNR for the whole GOP, and the rate on the horizontal axis corresponds to the average number of kbits per frame decoded from the scalable bit-stream. For the BIHA and TRAD schemes, the shaded areas on the left show the average number of bits spent on just the motion fields, and the dashed curve on the left shows the cost of coding breakpoints (almost identical in the BIHA and TRAD scheme).

Table 5.3 BD-PSNR and BD-Rate gains of the proposed BIHA scheme compared to TRAD_NOBREAKS on sequences affected by motion blur. Background (BG) and foreground (FG) motion activity is summarized as **S**till, **T**ranslation, **A**cceleration, **R**otation, and **Z**oom

Sequence	Activity		Resolution	BD-PSNR (dB)	BD-Rate (%)
	BG	FG			
Baseball^{MBLUR}	A	1A, 1RA	640 × 480	3.85	−33.32
Beach^{MBLUR}	S	1ZA, 2A	640 × 480	2.54	−27.78
Space^{MBLUR}	S	3A	640 × 480	2.78	−33.72
Winter^{MBLUR}	S	5A	640 × 480	5.47	−41.70
Average	-	-	-	3.73	−34.13

One can observe that the R-D performance of the BIHA scheme is approaching the one of SHVC at higher bit-rates, and even outperforms SHVC on the two natural sequences; at lower bit-rates, SHVC performs better. In the next section, we provide a number of ways how the proposed scheme could be further improved. Even so, our highly scalable algorithm is competitive with SHVC at higher bit-rates.

5.6   Potential for Improvements

We emphasize that in the comparisons with SHVC, we are comparing a mature codec with a scheme that has much potential for optimization. In this section, we give a list of suboptimalities in the present scheme; improving them would likely have a positive impact on the compression performance of the BIHA scheme. There are a number of places where the proposed BOA-TFI – which forms the essential building block of the BIHA scheme – can be improved.

1. **Motion Discontinuity Measure** In a compression scenario, where breakpoints are used to conserve the sharp motion discontinuities on quantized motion, it is quite a natural choice to also use breakpoints in the temporal reasoning. However, breakpoints are a binary representation of motion discontinuities; in particular, it is not possible to distinguish between discontinuities on the trailing side (useful to handle disocclusions, see Sect. 4.3.1), and discontinuities on the leading side of moving objects, where double mappings arise (see Sect. 4.2.2);
2. **Geometrical Consistency** Three motion field inversions are required to obtain the motion fields $\hat{M}_{b \to a}$ and $\hat{M}_{b \to c}$, which are used to predict the target frame. Each motion inversion can introduce errors, which results in the fact that the forward and backward pointing motion field are not guaranteed to be geometrically consistent;
3. **Texture Optimizations** We identify two cases where the presented scheme could be improved by applying texture optimizations:

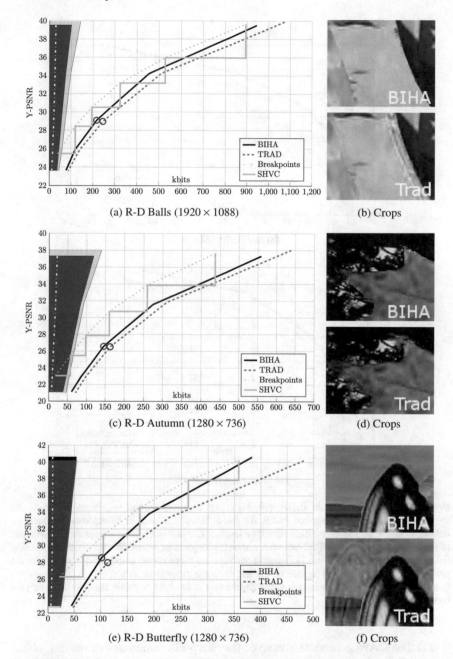

Fig. 5.10 R-D comparisons of BIHA with SHVC on synthetic sequences. We show the average per-frame bit-rate against Y-PSNR for various synthetic sequences, obtained using $T = 3$ temporal decomposition levels (GOP size $= 8$). The filled regions on the left show the average bit-rate spent on coding motion field data for the BIHA and TRAD schemes. (b/d/f) are crops of frames reconstructed at medium bit-rate (red circles in the R-D plots), for the BIHA and TRAD scheme, respectively. Note how the proposed BIHA scheme has much less ghosting artefacts than the traditional anchoring

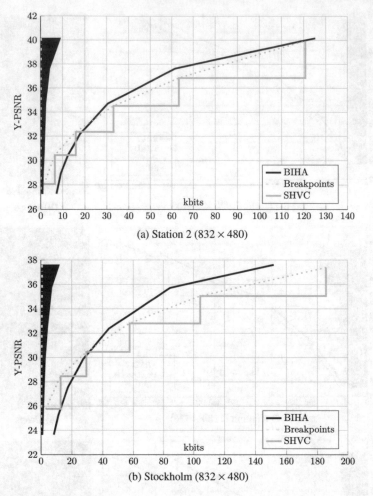

Fig. 5.11 R-D comparison of BIHA with SHVC on common natural sequences. We show the average per-frame bit-rate against PSNR for the **a** "Station 2" and **b** "Stockholm" sequences, obtained using $T = 3$ temporal decomposition levels (GOP size $= 8$). We compare the performance of the BIHA scheme with SHVC on two common test sequences, where the motion obtained using the optical flow estimator from [12] almost satisfies the hierarchical consistency required by the proposed method. The filled regions on the left show the average bit-rate spent on coding motion field data for the BIHA scheme; the white dashed line at the far left shows the cost of coding the breakpoints

(1) The CAW procedure preserves the sharp discontinuities at moving object boundaries in the mapped motion fields. The warped texture information thus exhibits sharp discontinuities as well, which can lead to a "cut-out" effect in the interpolated frames around moving objects.

(2) The estimated disocclusion information is used to switch from bidirectional to unidirectional prediction in regions that are only visible in one reference frame. In regions where the illumination drastically changes between the two reference frames, the transition between uni- and bidirectional prediction can be visible.

In a TFI scenario, improving the above-mentioned points is expected to have a positive impact on the visual quality of the interpolated frames. In a (scalable) video compression scenario, it is likely that it would improve the *prediction* performance, resulting in lower prediction residuals. In the next chapter, we propose modifications to BOA-TFI which address all the above-mentioned shortcomings. There are also improvements that are directly aimed at improving the compression performance.

1. **Coding of Inferred Motion Residuals**: Inferred motion residuals are another potential source of geometrical inconsistencies. Ideally, the prediction residuals of *inferred* motion fields are nonzero only in regions of disocclusions, where no inconsistencies can arise. However, in the current implementation, there is no guarantee that this is the case. In order to avoid such inconsistencies and avoid spending unnecessary bits on parts of the inferred motion residual where it should be zero, one could only code the residuals in regions where the disocclusion mask $\hat{S}_{c \to b} = 0$;
2. **Joint Coding of Motion Fields**: Currently, each motion field (residual) of each frame at each level is coded independently. Motion fields could be coded more efficiently by jointly coding the motion information, since wavelet coefficients that are nonzero at a coarse temporal level are likely to be nonzero at finer temporal scales; the same reasoning applies for the horizontal and vertical components of motion.

5.7 Summary

In this chapter, we presented a novel paradigm for anchoring motion fields employed in video compression. The proposed *bidirectional hierarchical anchoring (BIHA)* of motion fields at *reference frames* has some major advantages compared to the traditional way of anchoring motion fields at target frames. Every motion field involved in motion-compensated temporal prediction is warped from reference to target frames – a process during which we readily observe disocclusions; this valuable information traditionally has to be communicated as side information. Furthermore, the motion fields we compute for temporal prediction warp texture in a *geometrically consistent* manner, even if the motion is quantized. The analytical model developed in Sect. 5.3 provides insight into the relative importance and hence the weights to be assigned to the different spatio-temporal texture and motion field subbands. To further improve the scalability attributes of the BIHA scheme, we propose a *hierarchical spatio-temporal breakpoint induction (HST-BPI)* scheme to induce breakpoints from coarse to fine spatio-temporal levels. The fundamental building block of the BIHA scheme is the BOA-TFI method presented in Sect. 4.4 of the previous chapter. As such, improving the TFI scheme can be expected to have a positive impact on the coding performance of the BIHA scheme; this is the topic of the next chapter.

References

1. D. Rüfenacht, R. Mathew, D. Taubman, Hierarchical anchoring of motion fields for fully scalable video coding, in *IEEE International Conference on Image Processing* (2014) (cited on pages vii, 77)
2. D. Rüfenacht, R. Mathew, D. Taubman, Bidirectional hierarchical anchoring of motion fields for scalable video coding, in *IEEE 18th International Workshop on Multimedia Signal Processing* (2014) (cited on pages vii, 77)
3. D. Rüfenacht, R. Mathew, D. Taubman, A novel motion field anchoring paradigm for highly scalable wavelet-based video coding. IEEE Trans. Image Proc. **25**(1), 39–52 (2016) (cited on pages vii, 77)
4. T. Wiegand, G.J. Sullivan, G. Bjøntegaard, A. Luthra, Overview of the H.264/AVC video coding standard. IEEE Trans. Circ. Syst. Video Tech. **13**(7), 560–576 (2003) (cited on pages 1, 15, 19, 21, 77)
5. G.J. Sullivan, J.-R. Ohm, W.-J. Han, T. Wiegand, Overview of the high efficiency video coding (HEVC) standard. IEEE Trans. Circ. Syst. Video Tech. **22**(12), 1649–1668 (2012) (cited on pages 1, 15, 22, 77, 95)
6. A. Secker, D. Taubman, Lifting-based invertible motion adaptive transform (LIMAT) framework for highly scalable video compression. IEEE Trans. Image Proc. **12**, 1530–1542 (2003) (cited on page 83)
7. A. Mavlankar, S.-E. Han, C.-L. Chang, B. Girod, A new update step for reduction of PSNR fluctuations in motion-compensated lifted wavelet video coding, in *IEEE International Workshop on Multimedia Signal Processing* (2005) (cited on page 83)
8. R. Mathew, D. Taubman, P. Zanuttigh, Scalable coding of depth maps with R-D optimized embedding. IEEE Trans. Image Proc. **22**(5), 1982-1995 (2013) (cited on pages 3, 5, 33, 34, 36, 91, 93, 96)
9. D. Taubman, High performance scalable image compression with EBCOT. IEEE Trans. Image Proc. **9**(7), 1158–1170 (2000) (cited on pages 14, 91)
10. D. Taubman, M.W. Marcellin, *JPEG2000: Image Compression Fundamentals, Standards, and Practice* (Kluwer Academic Publishers, Boston, 2002) (cited on pages 2, 9, 14, 93, 96)
11. P. Helle, H. Lakshman, M. Siekmann, J. Stegemann, T. Hinz, H. Schwarz, D. Marpe, T. Wiegand, A scalable video coding extension of HEVC, in *Proceedings of the IEEE Data Compression Conference* (2013) (cited on pages 2, 25, 39, 95)
12. L. Xu, J. Jia, Y. Matsushita, Motion detail preserving optical flow estimation, in *IEEE Transactions on Pattern Analysis and Machine Intelligence* (2012), pp. 1744–1757 (cited on pages 43, 44, 49, 73, 95, 102, 128, 130, 131, 134, 137, 139, 159, 185-187)
13. D. Rüfenacht, R. Mathew, D. Taubman, Motion blur modelling for hierarchically anchored motion with discontinuities, in *IEEE International Workshop on Multimedia Signal Processing* (2015) (cited on pages vii, 99)
14. J. Wulff, M.J. Black, Modeling blurred video with layers. Eur. Conf. Comput. Vis. **8694**, 236–252 (2014) (cited on pages 43, 99)

Chapter 6
Forward-Only Hierarchical Anchoring (FOHA) of Motion

The *bidirectional hierarchical anchoring (BIHA)* of motion presented in the last chapter constitutes a fundamental change in the way motion is anchored and employed in a video compression system. Anchoring motion at reference-frames might appear counter-intuitive, since the motion information has to be mapped to target frames in order to serve as prediction reference. However, as shown in the last two chapters, this change of motion anchoring has a number of key advantages over the traditional anchoring of motion. First, motion information at finer temporal levels can be "recycled" from coarser levels, via the motion scaling operation. Second, during the motion mapping process, disoccluded regions are readily observed; this valuable information has to be explicitly communicated in a traditional anchoring scheme.

The framework used to map motion from reference to target frames essentially performs TFI. We have shown that the fundamental building block of the BIHA scheme, which we call BOA-TFI, is able to produce state-of-the-art TFI results. In Sect. 5.6, we identified a number of potential improvements of the BOA-TFI scheme, which are likely to have a positive impact on compression performance as well. In this chapter, we augment BOA-TFI with the following contributions, which address the key issues identified in the BOA-TFI scheme:

1. **Motion Discontinuity Measure** We propose a *disocclusion and folding likelihood map (DFLM),* which improves the robustness of the motion inversion process, in particular in regions of complex geometry;
2. **Geometrical Consistency** We change the direction of motion field inference to a *forward* inference. This simplifies the motion anchoring to a forward-only anchoring, which leads to further improved geometrical consistency of the bidirectionally interpolated target frames, while at the same time reducing the computational complexity;

© Springer Nature Singapore Pte Ltd. 2018
D. Rüfenacht, *Novel Motion Anchoring Strategies for Wavelet-Based Highly Scalable Video Compression*, Springer Theses,
https://doi.org/10.1007/978-981-10-8225-2_6

3. **Texture Optimizations** Two effective *texture optimizations* are proposed that selectively improve problematic regions of the interpolated texture data, which improves the interpolated frames both visually as well as quantitatively.

We present the motion-divergence based DFLM in, and highlight its advantages over the breakpoint-based motion discontinuity measure employed in BIHA. In Sect. 6.2, we introduce the FOHA-TFI scheme, and show the required changes that have to be made to the motion warping process. The combination of a more robust motion discontinuity measure with the forward-only motion anchoring improves the quality of the warped motion fields that are used as a prediction references when compared to BOA-TFI. While the better motion fields have a positive impact on the reconstructed frames, the interpolated frames exhibit sharp discontinuities around moving objects; this "cut-out" effect is a consequence of the sharp discontinuities that are present in the mapped motion fields. Furthermore, artificial boundaries can appear in regions where the occlusion-aware scheme switches from bidirectional to unidirectional prediction; these visually disturbing "transition boundaries" are most visible in regions where the illumination changes between the two prediction references. In Sect. 6.3, we propose two texture optimizations that specifically target these regions, without affecting other regions that are expected to be correctly predicted.

In Sect. 6.4, we extensively evaluate the TFI performance of the FOHA-TFI scheme, and compare it with state-of-the-art TFI schemes from the literature, as well as with BOA-TFI. While the focus of this chapter is on improving the TFI performance, we outline a highly scalable video compression scheme in Sect. 6.5 which is based on FOHA-TFI, which we call FOHA, and briefly discuss potential advantages over the BIHA scheme.

6.1   Disocclusion and Folding Likelihood Map (DFLM)

In the context of highly scalable video compression, where we use breakpoints to drive the breakpoint-adaptive DWT to efficiently compress motion fields, it is quite a natural choice to also use breakpoints as the basis for reasoning about motion discontinuities. However, the breakpoints themselves are a binary description that contains only some of the information related to motion field discontinuities. In this section, we describe an alternate approach that works directly with the motion data, using *motion divergence* as a soft, signed indicator of disocclusion and folding of the motion field [1].

We use $u(x, y)$ and $v(x, y)$ to denote the horizontal and vertical component of a motion field $M_{i \to j}$. The divergence of the motion field is defined as

$$div(M_{i \to j}) = \nabla \cdot M_{i \to j} = \frac{\delta u}{\delta x} + \frac{\delta v}{\delta y}. \tag{6.1}$$

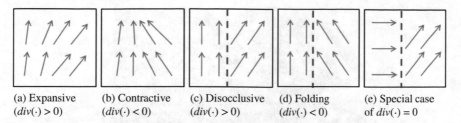

(a) Expansive   (b) Contractive   (c) Disocclusive   (d) Folding   (e) Special case
$(div(\cdot) > 0)$ $(div(\cdot) < 0)$ $(div(\cdot) > 0)$ $(div(\cdot) < 0)$ of $div(\cdot) = 0$

Fig. 6.1 Motion vector constellations for divergence analysis. **a** positive divergence indicates expansive motion, whereas **b** negative divergence is caused by contractive motion. In the presence of motion discontinuities (dashed vertical line), **c** positive divergence is indicative of disocclusion, whereas **d** negative divergence indicates folding. **e** depicts a particular case where the divergence is zero in the presence of a motion discontinuity; in this case, no disocclusion or folding arises. **a–d** adapted from [1]

Let us next consider the discrete case. Letting $u[m, n]$ and $v[m, n]$ be the horizontal and vertical component of a *discrete* motion field, the divergence at pixel location $\mathbf{m} = [m, n]$ can be computed as

$$div(M_{i \to j}[\mathbf{m}]) = u[m+1, n] - u[m-1, n] + v[m, n+1] - v[m, n-1]. \quad (6.2)$$

In Fig. 6.1, we show different configurations of motion vectors, together with the corresponding sign of the divergence value. Divergence can be used to distinguish geometric *expansion* $(div(\cdot) > 0)$ and *contraction* $(div(\cdot) < 0)$; this is visualized in Fig. 6.1a/b. Figure 6.1c/d show cases where motion discontinuities (vertical dashed lines in the figure) give rise to occlusions; more precisely, positive divergence is indicative of *disocclusion* (Fig. 6.1c), whereas negative divergence (i.e., convergent motion) is indicative of *folding* (Fig. 6.1d). We use the fact that the derivative of the divergence is nonzero at motion discontinuities to distinguish disocclusion and folding from largely expanding or contracting regions, respectively. Figure 6.1e shows a special case of a motion field with motion discontinuity, where the divergence is zero. We point out that the fact that the divergence is zero implies that such motion discontinuities do not give rise to disocclusion or folding. In other words, reasoning about divergence allows us to only consider motion discontinuities where disocclusion or folding happens.

In order to account for the fact that on real data the exact location of the motion discontinuity is unknown, we convert the motion divergence map to a *disocclusion and folding likelihood map (DFLM)*. The DFLM for a motion field $M_{i \to j}$, denoted as $\hat{D}_{i \to j}$, is obtained by applying a Gaussian blur to the motion divergence field:

$$\hat{D}_{i \to j}[\mathbf{m}] = (div(M_{i \to j}) * h)[\mathbf{m}], \quad (6.3)$$

where $h[\mathbf{m}]$ is a two-dimensional Gaussian kernel. While one could envision an adaptive filter size, we found that a fixed size for $h[\mathbf{m}]$ (7×7) works well on a wide

(a) Bandage 1 $M_{3 \to 4}$

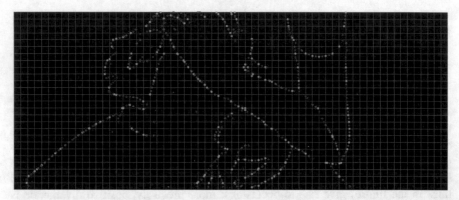

(b) Breaks

(b) DFLM

Fig. 6.2 Motion discontinuity representations estimated on the motion field in **a**, using **b** break-points, and **c** the DFLM proposed in this chapter. Breakpoints in the same cell (of any colour) are connected together to form discontinuity line segments, which serve as a binary representation of motion discontinuities. The DFLM can be used to distinguish between leading (red) and trailing side (cyan) of moving objects, which is very valuable in the identification of foreground and background objects

range of sequences. For consistency of notation, we will use $\hat{D}_{i \rightarrow j}(\mathbf{x})$ whenever we refer to accessing a *continuous* location of the DFLM, which is obtained using bilinear interpolation of the discrete DFLM.

Figure 6.2 shows an example estimated breakpoint field and DFLM. For viewing purposes, we show a subsampled version of the breakpoint field; the different colours of the dots indicate vertices that become alive at that level (red), vertices that became alive at a coarser level (orange), and spatially induced breaks (magenta); For the purpose of inducing motion discontinuities, they are treated the same; that is, whenever a cell contains two breakpoints, they are connected together to form a discontinuity line segment.

For the DFLM, we use cyan for divergent regions, which arise on the *trailing* side of moving objects, and red for convergent (i.e., negative divergence) regions, which arise on the leading side of objects in motion. Visual inspection of Fig. 6.2 shows how breakpoints only inform about the *presence* of a motion discontinuity, whereas the *signed* DFLM further allows us to deduce the type (e.g., divergent or convergent) of motion, as well as magnitude of the divergence.

We now provide some more motivation behind the use of motion divergence as a measure of disocclusion and folding. Let us first consider one rigid object moving on top of another rigid object; in this case, the divergence at the motion boundary measures the relative motion difference of the two objects in the direction *normal* to the boundary, which is exactly where disocclusion and folding arises. In the case of non-rigid objects, disocclusion and folding involves motion whose projection onto the boundary normal is *discontinuous* at the boundary; hence, the *continuous* derivative in that direction is infinite, which swamps any *finite* contribution of non-rigid transformations to the divergence. Of course, the discrete divergence approximation we use does not produce infinities; nonetheless the above argument essentially remains intact, so that large positive divergence (e.g., $div(M_{i \rightarrow j}) > +1$) is indicative of disocclusion, whereas $div(M_{i \rightarrow j}) < -1$ is indicative of folding.

6.2 Forward-Only Anchoring of Motion

In this section, we present the FOHA-TFI scheme, and contrast it with the BOA-TFI scheme introduced in Sect. 4.4. Figure 6.3 depicts the motion anchoring[1] used in the FOHA-TFI scheme, as well as the BOA-TFI scheme. In the same way as BOA-TFI forms the main building block of the BIHA scheme for highly scalable video compression, FOHA-TFI forms the essential building block of what we refer to as forward-only hierarchical anchoring (FOHA) scheme for video compression.

From the figure, one can see that the anchoring (and therefore direction) of the *inferred* motion field (dashed arrows in the figure) is flipped. As we will see in this section, this has a number of advantages over the BOA-TFI scheme. In fact, the

---

[1]The term motion anchoring is more meaningful in a compression scenario, where the anchoring refers to motion fields that are *coded*.

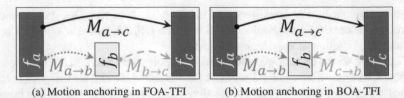

(a) Motion anchoring in FOA-TFI   (b) Motion anchoring in BOA-TFI

Fig. 6.3 Comparison between the motion anchoring of **a** FOHA-TFI and **b** BOA-TFI. One can see that the direction of the *inferred* motion field (dashed arrow) is reversed. While not apparent in the figure, this change guarantees geometrical consistency of the mapped motion fields that serve as prediction references

change of anchoring of the inferred motion field increases the geometrical consistency of the mapped motion fields, while at the same time reducing the computational complexity of the frame interpolation process by roughly a factor of 3. We explain why this is in more detail in Sect. 6.2.2; the reduced computational complexity is also confirmed on a comprehensive experimental evaluation in Sect. 6.4.3.

The change of motion anchoring and the new motion discontinuity measure (introduced in the last section) are two separate changes. However, they address different shortcomings of BOA-TFI, and therefore we decided to present their joint impact. Consequently, we use FOHA-TFI to refer to the scheme that uses the forward-only motion anchoring and employs the DFLM as motion discontinuity measure; similarly, BOA-TFI refers to the bidirectional anchoring of motion which uses breakpoints as motion discontinuity measure.

We give a high-level overview of the FOHA-TFI scheme in Sect. 6.2.1. FOHA-TFI necessitates motion fields to be mapped from reference to target frames, as was the case for BOA-TFI. The main ideas that are used to disambiguate double mappings, as well as to assign sensible motion in disoccluded regions, remain the same as the ones used in the BOA-TFI method. We then present the required changes to the TFI scheme. In particular, we detail the modifications needed to incorporate the DFLM as alternate motion discontinuity measure in Sect. 6.2.3. In Sect. 6.2.4, we present how disoccluded regions are detected and handled in the FOHA-TFI framework. Lastly, we compare FOHA-TFI with BOA-TFI in Sect. 6.2.5, and highlight the joint advantages of FOHA-TFI and the DFLM.

6.2.1 FOA-TFI Overview

Guided by Fig. 6.4, we give an overview of FOHA-TFI, on the example of frame upsampling by a factor of 2; arbitrary upsampling factors can be obtained by choosing an appropriate scaling factor α. The first step of the proposed method consists of estimating motion discontinuities by computing the DFLM on the input motion fields $M_{a\to c}$ and $M_{c\to e}$. Next, we *scale* the "parent" motion field $M_{a\to c}$ by a factor of 0.5 to obtain an estimate of $\hat{M}_{a\to b}$, under constant velocity assumption. $\hat{M}_{a\to b}$,

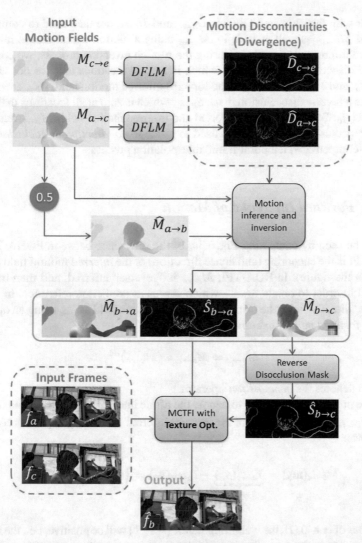

Fig. 6.4 Overview of FOHA-TFI. Input to the scheme are a motion field $M_{a\rightarrow c}$ (and $M_{c\rightarrow e}$, only used for motion discontinuities), and the two reference frames f_a and f_c. In the first step, we compute the (signed) divergence of the input motion fields $M_{a\rightarrow c}$ and $M_{c\rightarrow e}$ to obtain $\hat{D}_{a\rightarrow c}$ and $\hat{D}_{c\rightarrow e}$, respectively. $M_{a\rightarrow c}$ is then *scaled* by a factor of 0.5 to obtain $\hat{M}_{a\rightarrow b}$. Using information about motion discontinuities (i.e., discontinuities displace with foreground objects), $\hat{M}_{a\rightarrow b}$ is inverted to obtain $\hat{M}_{b\rightarrow a}$; at the same time, we *infer* $\hat{M}_{b\rightarrow c}$. During this process, we compute a forward disocclusion mask $\hat{S}_{b\rightarrow a}$; the reverse disocclusion mask $\hat{S}_{b\rightarrow c}$ is computed using reasoning on the inferred motion field $\hat{M}_{b\rightarrow c}$. Together, they are used to guide the bidirectional motion-compensated TFI process to obtain the interpolated frame $\hat{f}_b$

together with the estimated DFLMs $\hat{D}_{a \to c}$ and $\hat{D}_{c \to e}$, are then used to compute the *inverted* $\hat{M}_{b \to a}$, as well as to *infer* $\hat{M}_{b \to c}$ using a motion operation we refer to as "forward inference" of motion. During the motion inversion process, we compute a forward disocclusion mask $\hat{S}_{b \to a}$, which is zero at locations that are not visible in either f_a, and 1 elsewhere. From the forward inferred motion field $\hat{M}_{b \to c}$, we further compute a reverse disocclusion mask $\hat{S}_{b \to c}$, which is zero at all locations that are not visible in f_c. Together with the inverted and inferred motion fields $\hat{M}_{b \to a}$ and $\hat{M}_{b \to c}$, these disocclusion masks are then used to guide the *bidirectional, occlusion-aware* motion-compensated temporal frame interpolation process.

6.2.2   Forward Inference of Motion

As can be seen by comparing Fig. 6.3a, b, the difference between FOHA-TFI and BOA-TFI is the anchoring (and hence direction) of the *inferred* motion field (dashed arrow in the figure). In BOA-TFI, $\hat{M}_{c \to b}$ is "reverse" inferred, and then has to be inverted in order to obtain $\hat{M}_{b \to c}$, which serves as prediction reference. In FOHA-TFI, we "directly" infer the (forward pointing) motion field $\hat{M}_{b \to c}$ using an operation we call *forward motion inference*:

$$\hat{M}_{b \to c} = M_{a \to c} \circ (\hat{M}_{a \to b})^{-1}, \tag{6.4}$$

where $\circ$ denotes the *composition* operator.

More precisely, using $T_{i \to j}$ to denote the affine interpolated motion field obtained from $M_{i \to j}$, each location $\mathbf{m}_b$ of the forward inferred motion $\hat{M}_{b \to c}[\mathbf{m}_b]$ is computed as follows:

$$\hat{M}_{b \to c}[\mathbf{m}_b] = \underbrace{T_{a \to c}(\mathbf{x}_a)}_{= \frac{1}{\alpha} \cdot T_{a \to b}(\mathbf{x}_a)} - T_{a \to b}(\mathbf{x}_a) = \left(\frac{1}{\alpha} - 1 \right) T_{a \to b}(\mathbf{x}_a). \tag{6.5}$$

For values of $\alpha \in]0, 1[$, the weighting factor $\left(\frac{1}{\alpha} - 1 \right)$ will be positive, i.e., the inferred motion points in the same direction as the scaled motion $T_{a \to b}(\mathbf{x}_a)$, hence the term *forward* inference.

As can be seen in (6.4), the motion *inference* process involves the *inversion* of $\hat{M}_{a \to b}$; what this means is that the motion *inversion* and *inference* are performed *jointly*, leading to geometrical consistency of the motion fields $\hat{M}_{b \to a}$ and $\hat{M}_{b \to c}$, which are used to interpolate the target frame f_b. In contrast, the BOA-TFI framework we presented in Sect. 4.4 necessitates multiple motion field inversions, and geometrical consistency – in particular around motion discontinuities – is not guaranteed; we discuss the consequences of this in Sect. 6.2.5, and refer to the visual results in Fig. 6.9.

6.2.3 Resolving Double Mappings Using DFLM

In the FOHA-TFI scheme, the inverted and the forward inferred motion fields $\hat{M}_{b \to a}$ and $\hat{M}_{b \to c}$ are obtained using the same CAW procedure as the one used in BOA-TFI presented in Sect. 4.2. In short, triangles of affine motion are formed in the reference frame and mapped to the target frame. During this mapping process, as observed earlier, multiple triangles might overlap at location $\mathbf{m}_b$ in the target frame f_b. In other words, there are (at least) two locations $\mathbf{x}_a^1$ and $\mathbf{x}_a^2$ in f_a, which are mapped by $T_{a \to b}$ to the same location $\mathbf{m}_b$ in f_b, i.e.,

$$\mathbf{x}_a^1 + T_{a \to b}(\mathbf{x}_a^1) = \mathbf{x}_a^2 + T_{a \to b}(\mathbf{x}_a^2) = \mathbf{m}_b. \qquad (6.6)$$

We remind the reader that for BOA-TFI, we proposed a HST-BPI procedure to warp motion discontinuity information – represented using breakpoints – to the target frame f_b, and double mappings were resolved using the observation that motion discontinuities travel with the foreground object. The DFLM does not lend itself nicely to be mapped to the target frame; instead, we make all the reasoning about motion discontinuities to identify the foreground moving between the reference frames. Guided by Fig. 6.5, we now provide a more detailed description of how double mappings can be resolved using DFLM.

The figure shows (colour-coded) motion fields for two consecutive frames (f_a and f_c) of the input sequence, where a foreground object moves on top of a static background.[2] The grey region in the top row of Fig. 6.5 outlines the true inverse motion field at frame f_b, which is not known to the procedure. The key idea behind the proposed method is that the two points $\mathbf{x}_a^1$ and $\mathbf{x}_a^2$ that create the double mapping in f_b must be separated in the reference frame f_a; moreover, along the line connecting the two points in f_a, denoted as l, there must be a region of *convergent* (i.e., negative divergence) motion.

If we map each point on line l from frame f_a to frame f_b (see bottom row of Fig. 6.5), using $T_{a \to b}$, we expect to see a discontinuous jump, which corresponds to the point of maximum convergence in frame f_a. We denote the points on either side of the discontinuous jump as B_a^1 and B_a^2. The location of these points mapped to f_c is

$$B_c^p = B_a^p + T_{a \to c}(B_a^p), \; p \in \{1, 2\}. \qquad (6.7)$$

The importance of these points is that one of the B_c^ps is expected to fall into a region of similar (i.e., negative) divergence, whereas the other is not; the foreground motion is the motion of the point which maps into the region of large(r) negative divergence. That is, the motion of the *inverted* motion field $\tilde{M}_{b \to a}$ at the double mapped location $\mathbf{m}_b$ is

---

[2]The example uses a static background for ease of explanation; we note that the method remains valid for any combination of moving foreground and background objects.

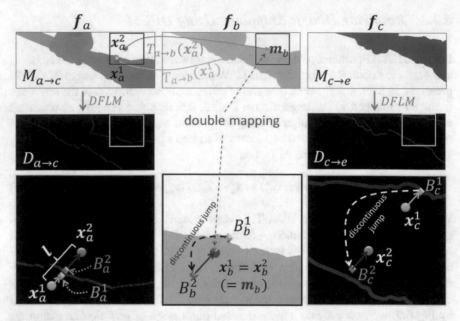

Fig. 6.5 Illustration of how double mappings are resolved using the DFLM as motion discontinuity representation. When mapped from f_a to f_b using $T_{a \to b}$, $\mathbf{x}_a^1$ and $\mathbf{x}_a^2$ map to the same location $\mathbf{m}_b$ in f_b. We use the DFLMs $\hat{D}_{a \to c}$ and $\hat{D}_{c \to e}$ as motion discontinuity measures (see sec:dflm) to identify the foreground motion vector. We search for the point of maximum negative divergence along the line I, formed by connecting $\mathbf{x}_a^1$ and $\mathbf{x}_a^2$. Let B_a^1 and B_a^1 be the two points on the line slightly closer to $\mathbf{x}_a^1$ and $\mathbf{x}_a^2$, respectively. When B_a^1 and B_a^1 are mapped to the other reference frame f_c, the one which maps into a region of larger negative divergence (red) identifies the foreground motion; here, $T_{a \to b}(\mathbf{x}_a^1)$ is the foreground motion, since B_c^1 falls into a region of larger negative

$$
\hat{M}_{b \to a}[\mathbf{m}_b] = \begin{cases} -T_{a \to b}(\mathbf{x}_a^1) & \hat{D}_{c \to e}(B_c^1) < min(\hat{D}_{c \to e}(B_c^2), -\Theta) \\ -T_{a \to b}(\mathbf{x}_a^2) & \hat{D}_{c \to e}(B_c^2) < min(\hat{D}_{c \to e}(B_c^1), -\Theta) \, , \\ \hat{M}_{b \to a}^{old}[\mathbf{m}_b] & \text{otherwise} \end{cases}
\tag{6.8}
$$

where $\Theta > 0$ is a threshold that guarantees that only motion is selected that falls into a region of negative divergence. In the rare case where $min(\hat{D}_{c \to e}(B_c^1), \hat{D}_{c \to e}(B_c^2)) > -\Theta$, the previously assigned motion $\hat{M}_{b \to a}^{old}[\mathbf{m}_b]$ is kept. Combining (6.5) and (6.8), the motion of the *forward inferred* motion field $\hat{M}_{b \to c}$ at location $\mathbf{m}_b$ is readily assigned as

$$
\hat{M}_{b \to c}[\mathbf{m}_b] = \left(1 - \frac{1}{\alpha}\right) \hat{M}_{b \to a}[\mathbf{m}_b].
\tag{6.9}
$$

An important advantage of the DFLM is that it allows us to approximately halve the discontinuity boundaries that have to be considered compared to the (unsigned) discontinuity representation induced from breakpoints. This proves particularly useful

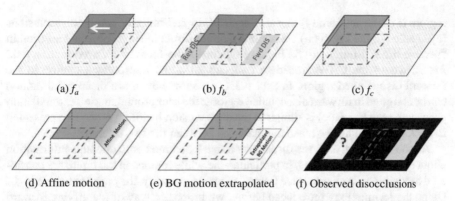

(a) f_a (b) f_b (c) f_c

(d) Affine motion   (e) BG motion extrapolated   (f) Observed disocclusions

Fig. 6.6 Illustration of forward and reverse disocclusions. The rectangle moves from right to left between the **a** the left reference frame f_a and **c** the right reference frame f_c, on top of static background. As shown in **b**, forward disocclusions (yellow) are regions that get uncovered as objects travel *forward in time*. Such regions should only be predicted from f_c; however, as shown in **d**, the *affine* interpolated motion assigned by the CAW procedure is "non-physical", and should be replaced by extrapolated background motion (**e**). Reverse disocclusions (cyan) are regions that get uncovered as objects transition in "reverse time". In such regions, valid motion is assigned. However, unlike forward disocclusions, the procedure does not readily observe reverse disocclusions; **f** the white regions are considered as visible in both reference frames, whereas in reality, they are not visible in f_c

in regions of complex geometry, as well as for thin moving objects; examples can be seen in Fig. 6.9.

6.2.4 Handling of Forward and Reverse Disocclusions

One key property of the motion field inversion operation presented in Sect. 4.2 is that the inversion of a motion field $M_{i \to j}$ allows us to readily observe regions that get disoccluded between frames f_i and f_j. In the BOA-TFI scheme (see Fig. 6.3b), both the *scaled* motion field $\hat{M}_{a \to b}$ *and* the reverse inferred motion field $\hat{M}_{c \to b}$ are anchored at reference frames. In order to serve as prediction references, they both have to be inverted, and hence disoccluded regions are observed for both reference frames. In FOHA-TFI, on the other hand, the forward inferred motion field is anchored at the target frame. The consequence is that while FOHA-TFI readily observes regions that get disoccluded between f_a and f_b during the inversion of $\hat{M}_{a \to b}$, it does not observe which regions of f_b are not visible in f_c as part of the motion mapping process.

We use Fig. 6.6 to guide the ensuing discussion about how disoccluded regions are handled in FOHA-TFI. In the example of the figure, a rectangle moves from right to left between frames f_a and f_c. We find it helpful to distinguish between *forward* and *reverse* disocclusions. As can be seen in Fig. 6.6b, forward disocclusions (yellow area) are regions that get disoccluded between frames f_a and f_b; they correspond to regions in the target frame f_b that are *not visible* in f_a. Similarly, reverse disocclusions (cyan areas in the Fig. 6.6b) are regions that get disoccluded as objects transition in "reverse time" from f_c to f_b. Figure 6.6d shows the forward inferred motion field $\hat{M}_{b \to c}$, where the CAW procedure assigned an affine interpolated motion in the forward disoccluded region. In Sect. 6.2.4, we show how more meaningful motion can be assigned in forward disoccluded regions; the background motion extrapolation procedure (see Fig. 6.6e) is similar to the disocclusion handling procedure presented in Sect. 4.3.1, but all the reasoning happens between the two reference frames.

As shown in Fig. 6.6f, the disocclusion mask obtained only contains information about forward disocclusions. In particular, the white regions in the figure are considered as visible in both reference frames, whereas in reality, they are not visible in f_c. Using the terminology introduced before, we propose a way of identifying forward disoccluded regions in Sect. 6.2.4.2.

6.2.4.1   Changing Motion in Forward Disoccluded Regions

Mapping triangles to frame f_b using $T_{a \to b}$ exposes regions that get forward disoccluded between frames f_a and f_b; as mentioned earlier, this information is useful to guide the bidirectional frame interpolation process as shown in (4.12), and we record it in a *forward disocclusion* mask $\hat{S}_{b \to a}$:

$$\hat{S}_{b \to a}[\mathbf{m}] = \begin{cases} 0 & \mathbf{m} \text{ disoccluded} \\ 1 & \text{otherwise} \end{cases}. \tag{6.10}$$

In regions where $\hat{S}_{b \to a}[\mathbf{m}] = 0$, the affine interpolated motion between background and foreground motion produced by the CAW method (see Fig. 6.6d) is inappropriate. In fact, regions that are disoccluded when going from frame f_a to f_b are likely to be visible in frame f_c; we therefore want to assign a motion that maps forward disoccluded pixels to the corresponding location in f_c. This can be realised by extrapolating background motion in such forward disoccluded regions (see Fig. 6.6e).

As it turns out, no explicit distinction between foreground and background motion needs to be made to achieve the desired result. Key to the proposed procedure of assigning "appropriate" motion in forward disoccluded regions is the use of motion discontinuities, which mark the end of the disoccluded region. As can be seen in Fig. 6.7a, the intersection of motion discontinuities with the triangle in frame f_a is somewhat arbitrary; in particular, mapping these intersections to f_b under the affine motion assumption is unlikely to produce a good estimate of the "true" location of the motion discontinuity. By contrast, using the affine model to transfer boundary locations from the stretched triangle in f_c to target frame f_b yields a much more reliable estimate of the "true" boundary location. To see this, observe that the contractive transform from f_c to f_a necessarily maps corresponding motion boundaries to an accuracy of less than one pixel separation. At half the frame separation, the mapped boundary discrepancy in frame f_b should then be no larger than half a pixel.

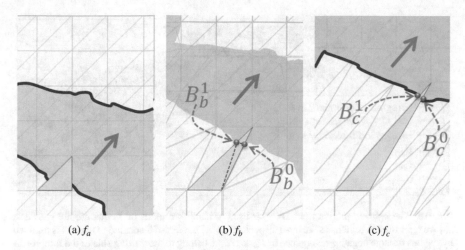

(a) f_a (b) f_b (c) f_c

Fig. 6.7 Illustration of the motion extrapolation method we employ in *forward disoccluded* regions. Since there is no motion field anchored at the target frame f_b, we use $\hat{D}_{a \to c}$ and $\hat{D}_{c \to e}$ to drive the motion extrapolation at frame f_b. Mapped by $T_{a \to c}$, two edges of the stretched "disoccluding" triangle are expected to intersect with regions of large divergence in $\hat{D}_{c \to e}$ of f_c (solid blue line). The point of maximum divergence along the two edges is mapped to the target frame f_b using the appropriately scaled affine motion; this gives a good estimate of the location of the motion discontinuity. The last step extrapolates the motion from the triangle vertices up to the motion discontinuities

Consequently, the procedure we propose to extrapolate motion in forward disoccluded regions finds the discontinuity boundaries in f_c, which are then mapped to the target frame f_b. To locate motion boundaries in f_c, we find the maximum DFLM value in $\hat{D}_{c \to e}$ along the three edges of the stretched triangle in frame f_c. We expect an isolated large value along two of the three edges in $\hat{D}_{c \to e}$; we call such edges *aligned* edges, since they are expected to connect motion vectors from two different moving objects.

Let B_c^p, $p \in \{0, 1\}$, denote the points of largest divergence along the two split edges of the disoccluding triangle. Ultimately, we are interested in the location of these discontinuities in the target frame f_b. Using $\mathbf{u}_{\text{aff}}(\cdot)$ to denote the affine function that describes motion $\hat{M}_{c \to a}$ over the triangle in f_c as obtained by the CAW procedure, the location of the discontinuity in the target frame f_b can be obtained as

$$B_b^p = B_c^p - 0.5\mathbf{u}_{\text{aff}}(B_c^p). \tag{6.11}$$

Connecting the two mapped B_b^ps splits the triangle in f_b into a triangle and a quadrilateral (see Fig. 6.7b).

The last step of the motion extrapolation procedure is to copy the motion from the vertices up to the motion discontinuities B_b^p; this effectively extrapolates the background motion in the background region of the disoccluded triangle. Similarly, in the (much smaller) foreground region, the foreground motion will be extrapolated.

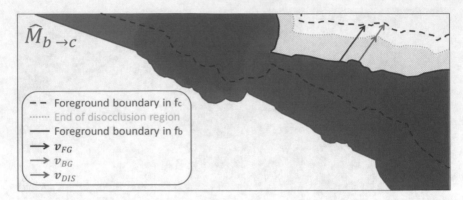

Fig. 6.8 The *inferred* motion field $\hat{M}_{b \to c}$ is used to identify regions in f_b that are not visible in f_c; we refer to such regions as "reverse disocclusions". The start of disoccluded regions is outlined by regions of large negative divergence in $\hat{D}_{b \to c}$; these belong to the trailing side of the foreground object as it moves (in reverse time) from f_c to f_b. $\mathbf{v}_{DIS}$, which points to the end of the disoccluded region, is defined by the relative difference in motion between the foreground ($\mathbf{v}_{FG}$) and background motion ($\mathbf{v}_{BG}$)

6.2.4.2 Detecting Reverse Disocclusions

In the proposed bidirectional prediction framework, we are also interested in regions of the target frame f_b which are not visible in frame f_c. In such regions of *reverse disocclusion*, the motion assigned to $\hat{M}_{b \to a}[\mathbf{m}]$ (and $\hat{M}_{b \to c}[\mathbf{m}]$) is *appropriate*; that is, it can be expected to map to a corresponding location in f_a, where it is visible. The challenge is, however, to know that they are not visible in f_c. In this section, we describe how a *reverse disocclusion* mask $\hat{S}_{b \to c}$ can be computed; we guide the description using Fig. 6.8, where a foreground object moves on top of a background that is moving in the opposite direction; that is, both foreground and background are in motion in this example.

The proposed procedure uses the *inferred* motion field $\hat{M}_{b \to c}$ and the DFLM $\hat{D}_{b \to c}$ to detect regions in f_b that are *not visible* in f_c. We observe that regions of large *negative* divergence in $\hat{D}_{b \to c}$ correspond to the trailing side of foreground objects as they (hypothetically) move (backwards in time) from frame f_c to f_b; this is the start of any reverse disoccluded region in $\hat{S}_{b \to c}$ (solid red line in Fig. 6.8). The "width" of the reverse disoccluded region can be approximated by a *disocclusion vector* $\mathbf{v}_{DIS}$, which points to the end of the disocclusion region (dotted green line in the figure):

$$\mathbf{v}_{DIS} = \mathbf{v}_{FG} - \mathbf{v}_{BG}, \tag{6.12}$$

where $\mathbf{v}_{FG}$ and $\mathbf{v}_{BG}$ are motion vectors from the fore- and background motion, respectively, as identified during the double mapping resolving procedure presented in Sect. 6.2.3. In practice, so as to obtain closed reverse disoccluded regions, we densely sample such disocclusion vectors in regions where $\hat{D}_{b \to c}$ is negative. The reverse disocclusion mask is then obtained as:

$$\hat{S}_{b\to c}[\mathbf{m}] = \begin{cases} 0 & \mathbf{m} \text{ covered by a } \mathbf{v}_{DIS} \\ 1 & \text{otherwise} \end{cases}. \tag{6.13}$$

6.2.5 Comparison Between BOA-TFI and FOA-TFI

In this section, we show the joint benefits of the forward-only anchoring *and* the DFLM to handle regions around moving objects on a concrete example. For this, we compare the temporal frame interpolation results obtained using BOA-TFI with FOHA-TFI. Figure 6.9 uses a frame from the "Market 2" sequence, which contains complex motion and a large variety of moving objects. The *estimated* disocclusion mask in Fig. 6.9a indicates difficult regions around moving object boundaries. Figure 6.9e, g and f, h show the backward and forward pointing motion fields obtained in the BOA-TFI and the FOHA-TFI scheme, respectively. The following two main observations can be made.

First, as a direct consequence of the fact that FOHA-TFI only involves the inversion of *one* motion field (as opposed to *three* inversions in BOA-TFI), the backward and forward pointing motion fields are much more consistent; in fact, in FOHA-TFI, it is guaranteed that $\hat{M}_{b\to c} = -\hat{M}_{b\to a}$, as can be seen by comparing the forward and backward pointing prediction fields in Fig. 6.9f, h.

The second observation is with respect to the new motion discontinuity measure. The DFLM is a *signed* measure of disocclusion and folding; it is positive on the trailing side of moving objects (blue in Fig. 6.9d), and negative on the leading side (red in Fig. 6.9d). By contrast, in the BOA-TFI approach, the discontinuity information derived from breakpoints does not carry enough information to explicitly distinguish between leading and trailing boundaries of moving objects. Therefore, the proposed FOHA-TFI is less likely to fail to resolve double mappings correctly. This is evidenced in the crops of the motion fields in Fig. 6.9e–h, where thin objects such as the hand and the arms of the girl, as well as the different apples falling out of the crate, create a multitude of motion discontinuities in close proximity, which lead to wrongly mapped motion in the BOA-TFI approach.

6.3 Texture Optimizations

The last step of FOHA-TFI is to use the mapped motion fields $\hat{M}_{b\to a}$ and $\hat{M}_{b\to c}$, together with disocclusion masks $\hat{S}_{b\to a}$ and $\hat{S}_{b\to c}$, to bidirectionally interpolate the target frame $\hat{f}_b$, as detailed for BOA-TFI in (4.12) of Sect. 4.4. Figure 6.10a shows an example of an interpolated frame; as can be seen in Fig. 6.10d, there can be visible

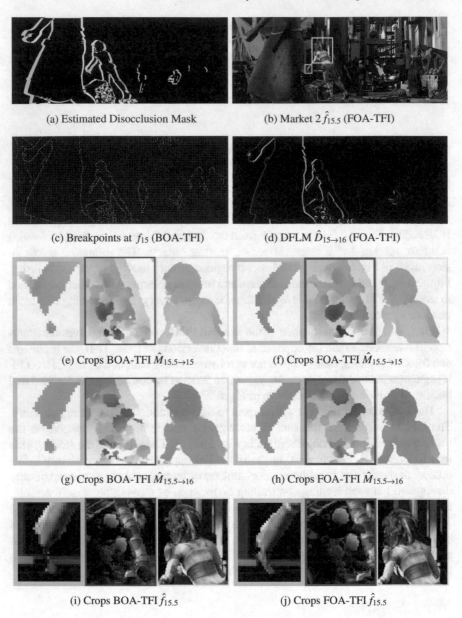

(a) Estimated Disocclusion Mask   (b) Market 2 $\hat{f}_{15.5}$ (FOA-TFI)

(c) Breakpoints at f_{15} (BOA-TFI)   (d) DFLM $\hat{D}_{15 \to 16}$ (FOA-TFI)

(e) Crops BOA-TFI $\hat{M}_{15.5 \to 15}$ (f) Crops FOA-TFI $\hat{M}_{15.5 \to 15}$

(g) Crops BOA-TFI $\hat{M}_{15.5 \to 16}$ (h) Crops FOA-TFI $\hat{M}_{15.5 \to 16}$

(i) Crops BOA-TFI $\hat{f}_{15.5}$ (j) Crops FOA-TFI $\hat{f}_{15.5}$

Fig. 6.9 Comparison of FOHA-TFI with BOA-TFI. **a** shows the union of the *estimated* forward and reverse disocclusion masks ($\hat{S}_{15.5 \to 15} \cup \hat{S}_{15.5 \to 16}$), where yellow and cyan indicate disoccluded regions in the previous and future reference frame, respectively; **b** shows the interpolated frame using the proposed FOHA-TFI method. **c** shows the estimated breakpoint map as employed in the BOA-TFI scheme; **d** shows the DFLM employed in the proposed FOHA-TFI method; **e–h** show crops of the backward and forward motion fields involved in the prediction of the target frame, where one can see the more accurate and consistent motion obtained using the FOHA-TFI scheme. **i** and **j** show the corresponding crops of the interpolated target frame

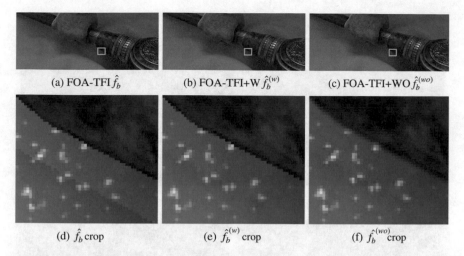

(a) FOA-TFI $\hat{f}_b$ (b) FOA-TFI+W $\hat{f}_b^{(w)}$ (c) FOA-TFI+WO $\hat{f}_b^{(wo)}$

(d) $\hat{f}_b$ crop   (e) $\hat{f}_b^{(w)}$ crop   (f) $\hat{f}_b^{(wo)}$ crop

Fig. 6.10 Example of the effects of the proposed texture optimizations. **a, d** show the bidirectionally predicted target frame $\hat{f}_b$, where there is a significant illumination change between the two reference frames in the disoccluded region. **b, e** show how the *selective wavelet coefficient attenuation* (+W) successfully creates a smoother transition from uni- to bidirectional prediction; **c, f** show the joint effect of the optical blur synthesis (+WO), which creates a smooth transition around moving objects, without smoothing any other part of the image

artefacts around moving objects. In the following, we propose two optimizations on the texture that mitigate these problems. It is worth noting that the proposed methods could be readily applied to the BOA-TFI procedure; however, since similar improvements can be expected in either method, we will focus on FOHA-TFI.

6.3.1 Selective Wavelet Coefficient Attenuation (SWCA)

At *disocclusion* boundaries, the upsampled frame interpolated using the occlusion-aware frame interpolation method of (4.12) can have problems; the sudden transition from uni- to bidirectional prediction can lead to artificial boundaries in places where the illumination changes between the two reference frames. We use Fig. 6.11 to illustrate this problem, where we focus on a disoccluded region of the "Bandage 1" sequence.

We observe that neither of the motion-compensated reference frames $f_{a \to b}$ and $f_{c \to b}$ is expected to contain such a transition boundary in the texture data. In the Fig. 6.11b, we show a crop of the HH_0 band, where no large wavelet coefficients are present around the transition boundary from uni- to bidirectional prediction (red circle in the figure). In Fig. 6.11c, we show the same region of the *blended* frame $\hat{f}_b$, where there are large coefficients around the transition boundary, which are visible as a sharp transition in the predicted target frame, as shown in Fig. 6.11d.

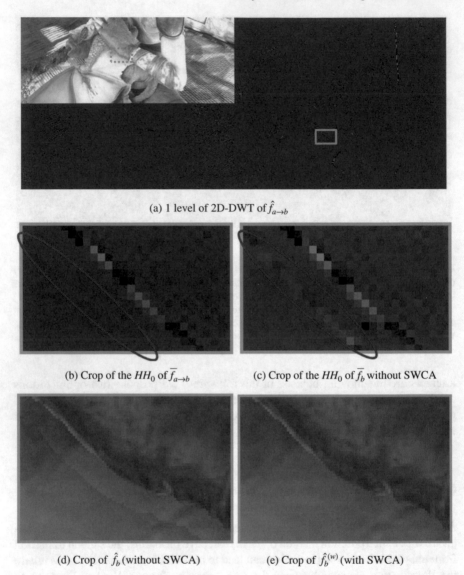

(a) 1 level of 2D-DWT of $\hat{f}_{a \to b}$

(b) Crop of the HH_0 of $\overline{f}_{a \to b}$ (c) Crop of the HH_0 of $\overline{f}_b$ without SWCA

(d) Crop of $\hat{f}_b$ (without SWCA)   (e) Crop of $\hat{f}_b^{(w)}$ (with SWCA)

Fig. 6.11 Illustrative example for the SWCA procedure. **a** shows a 1-level 2D-DWT of the warped reference frame $\hat{f}_{a \to b}$; **b** and **c** show crops of the HH_0 band of $\overline{f}_{a \to b}$ and $\overline{f}_b$ without SWCA, respectively. The red circle indicates the large magnitude coefficients introduced by averaging two warped frames with different illumination, which leads to a visible boundary at the transition boundary from uni- to bidirectional prediction in **d**. **e** shows the reconstructed frame where the proposed SWCA has attenuated the coefficients in the red circle, resulting in a much smoother transition from uni- to bidirectional prediction

Figure 6.11e shows the effect of the proposed SWCA, which is obtained by limiting the magnitude of the wavelet coefficients in the target frame $\hat{f}_b$, based on the wavelet decomposition of the motion compensated reference frames. We use the 5/3 wavelet transform, but the approach could be used with other (non-wavelet) transforms. For our work, the wavelet transform is particularly interesting for its useful connections with coding. The advantage of the proposed method over simple smoothing in the spatial domain is that it effectively realizes an adaptive smoothing filter size depending on the visibility of the transition boundary.

We use $\bar{f}_j$ to denote the 2D-wavelet decomposition of frame f_j, and use $\bar{f}_j[\mathbf{k}]$ to access a specific wavelet coefficient $\mathbf{k}$, where $\mathbf{k}$ collects information about level, subband, and spatial position in the transform. Furthermore, let $\check{S}_{i \to j}$ be a rearranged disocclusion mask $\hat{S}_{i \to j}$, so that the locations $\mathbf{k}$ in $\check{S}_{i \to j}[\mathbf{k}]$ correspond to the appropriate locations $\mathbf{m}$ in $\hat{S}_{i \to j}[\mathbf{m}]$. Then, we define

$$\tau[\mathbf{k}] = \max \left(\check{S}_{b \to a}[\mathbf{k}] | \bar{f}_{a \to b}[\mathbf{k}] |, \check{S}_{b \to c}[\mathbf{k}] | \bar{f}_{c \to b}[\mathbf{k}] | \right). \tag{6.14}$$

That is, $\tau[\mathbf{k}]$ represents the larger (visible) wavelet coefficient of the wavelet decomposition of the motion compensated left and right reference frames, evaluated at $\mathbf{k}$. Then, $\bar{f}_b$ is computed as follows:

$$\bar{f}_b^{(W)}[\mathbf{k}] = \begin{cases} \tau[\mathbf{k}] \cdot \text{sgn}(\bar{f}_b[\mathbf{k}]) & \tau[\mathbf{k}] \leq |\bar{f}_b[\mathbf{k}]| \\ \bar{f}_b[\mathbf{k}] & \text{otherwise} \end{cases}. \tag{6.15}$$

We then simply synthesize $\bar{f}_b^{(W)}$ to obtain $\hat{f}_b^{(W)}$. The effect this wavelet coefficient attenuation has in the spatial domain can be seen in Fig. 6.10e, where the sharp transition boundary from uni- to bidirectional prediction is much less visible; more examples are provided in Figs. 6.12 and 6.13.

What is particularly appealing about this selective wavelet coefficient attenuation is that it is applied *globally* to the entire frame, and there are no heuristics or parameters involved.

6.3.2 Optical Blur Synthesis

The SWCA presented in the previous section is effective at smoothing transitions from uni- to bidirectional prediction in regions with large illumination changes. Another visually disturbing artefact that can arise in the proposed scheme is that overly sharp transitions are created at moving object boundaries in the texture domain; this is because the inverted and inferred motion fields ($\hat{M}_{b \to a}$ and $\hat{M}_{b \to c}$, respectively), are discontinuous around moving object boundaries. This effectively cuts out the foreground object and pastes it in the target frame. In practice, the transition in the reference frames is smoother due to optical blur, which is an inevitable aspect

(a) Bandage 1 $\hat{f}_{4.5}$ (b) Crops FOA-TFI $\hat{f}_{4.5}$

(c) Crops FOA-TFI+W $\hat{f}_{4.5}^{(w)}$ (d) Crops FOA-TFI+WO $\hat{f}_{4.5}^{(wo)}$

(e) Crops $\hat{M}_{4.5\to4}$ (f) Crops $\hat{M}_{4.5\to5}$

Fig. 6.12 Different stages of the proposed FOHA-TFI method with texture optimization on a frame from the "Bandage 1" sequence. **a** shows the interpolated frame using FOHA-TFI+WO. **b–d** show crops of the interpolated frames obtained using the proposed FOHA-TFI method with no texture optimization, selective wavelet coefficient attenuation (+W), and additional optical blur synthesis (+WO), respectively. **e**, **f** show crops of the same region of the backward and forward pointing motion fields

of the imaging process. The wavelet-based attenuation strategy described above cannot resolve this problem because the unnaturally sharp discontinuities are expected to be present in both of the motion compensated reference frames.

We propose a simple yet effective way of synthesizing optical blur, which further improves the visual quality of the interpolated target frames, which uses the divergence of the target motion field $\hat{M}_{b\to c}$ as an indication of moving object boundaries. A Gaussian blur filter is applied to all pixels where the absolute value of the divergence of the motion field is larger than a certain threshold θ; the interpolated target frame with optical blur synthesis, denoted as $\hat{f}_b^{(WO)}$, is then obtained as:

$$\hat{f}_b^{(WO)} = \begin{cases} (\hat{f}_b^{(W)} * h)[\mathbf{m}] & |div(\hat{M}_{b\to c}[\mathbf{m}])| > \theta \\ \hat{f}_b^{(W)} & \text{otherwise} \end{cases}, \qquad (6.16)$$

(a) Kimono1 $\hat{f}_5$

(b) Crops FOA-TFI $\hat{f}_5$

(c) Crops FOA-TFI+W $\hat{f}_5^{(w)}$

(d) Crops FOA-TFI+WO $\hat{f}_5^{(wo)}$

(e) Crops $\hat{M}_{5\rightarrow4}$

(f) Crops $\hat{M}_{5\rightarrow6}$

Fig. 6.13 Different stages of the proposed FOHA-TFI method with texture optimization on a frame from the "Kimono1" sequence. **a** shows the interpolated frame using FOHA-TFI+WO. **b–d** show crops of the interpolated frames obtained using the proposed FOHA-TFI method with no texture optimization, selective wavelet coefficient attenuation (+W), and additional optical blur synthesis (+WO), respectively. **e, f** show crops of the same region of the backward and forward pointing motion fields

where $h[\mathbf{m}]$ is a two-dimensional Gaussian kernel. In our experiments, we found that $\theta = 5$ provides good results. Figure 6.10c, and perhaps more obviously Figs. 6.12d and 6.13d, provide examples of the benefits of the proposed optical blur synthesis. In the next section, we provide a more detailed discussion of the impact of the proposed texture optimizations.

6.3.3   Impact of Texture Optimizations

In this section, we assess the impact of the two proposed texture optimizations, namely the selective wavelet coefficient attenuation (FOHA-TFI+W) described in Sect. 6.3.1, as well as the *additional* motion blur synthesis (Sect. 6.3.2), referred to as FOHA-TFI+WO. We remind the reader that FOHA-TFI+W is used to smooth transitions from uni- to bidirectional prediction, where illumination changes between the two reference frames can add artificial high-frequency content. The optical blur synthesis aims at smoothing the transition from foreground to background texture at moving object boundaries. In the following, we first qualitatively and then quantitatively evaluate the impact of the proposed texture optimizations.

6.3.3.1   Visual Improvements

The proposed texture optimizations aim at improving the *visual* quality of the interpolated frames. Figure 6.12 illustrates the impact of the different steps of the proposed texture optimizations on the "Bandage 1" sequence from the Sintel dataset [2], which is a good example to illustrate the FOHA-TFI+W texture optimization, since there are significant illumination changes between the two reference frames. One can see in Fig. 6.12b how the FOHA-TFI without texture optimization contains visually disturbing high-frequency content at the transition point from uni- to bidirectional prediction. FOHA-TFI+W successfully creates a smoother transition by selectively attenuating the problematic large coefficients, without smoothing any of the correctly predicted parts of the frame.

Figure 6.13 shows an interpolated frame from the "Kimono1" sequence, where the motion is estimated using the optical flow method described in [3]. The "Kimono1" sequence is useful to illustrate the proposed optical blur synthesis. Without texture optimization (Fig. 6.13b), the sharp transitions in the inverted motion fields create a "cut-out" effect in the interpolated frame. This can be seen around the head of the woman; in addition, as can be seen in the motion field cops in Fig. 6.13e, f, there are different patches of background motion, which create subtle artificial transitions in the interpolated frame. For these regions, FOHA-TFI+W has little impact. The optical blur synthesis, which blurs the texture in regions of large *motion* divergence, is able to remove these artefacts and create a much smoother transition.

6.3.3.2   Quantitative Impact of Texture Optimizations

In order to quantitatively evaluate the impact of the two texture optimizations proposed in this chapter, we ran the TFI experiment on natural sequences as presented in Sect. 4.4.3; that is, for each of the twelve video sets (see Sect. A.3), we interpolated 10 frames, resulting in a total of 120 interpolated frames.

Table 6.1 Quantitative evaluation of the impact of the proposed textured optimizations on common natural test sequences. The table shows results for FOHA-TFI without texture optimizations (−), with enabled SWCA (+W), as well as additional optical blur synthesis (+WO). As an anchor point, we further replicate the results of the BOA-TFI [4] scheme (see Sect. 4.4). In parentheses (·), we show the difference between the Y-PSNR of FOHA-TFI+WO and the respective method we compare it to ("−" means that the proposed FOHA-TFI+WO performs better, "+" means worse performance). **Bold** indicates per-row best-performance

Sequence	BOA-TFI[a] [4]	Proposed FOHA-TFI[a]		
		−	+W	+WO
Cactus	33.63 (−0.68)	33.97 (−0.35)	34.13 (−0.18)	**34.31**
Kimono1	33.26 (−0.77)	33.65 (−0.38)	33.82 (−0.21)	**34.03**
Kimono2	40.94 (−0.69)	41.38 (−0.26)	41.39 (−0.24)	**41.63**
Rushhour	34.76 (−0.36)	34.95 (−0.17)	35.05 (−0.07)	**35.12**
Shields1	36.55 (−0.03)	36.35 (−0.24)	36.41 (−0.17)	**36.58**
Shields2	37.76 (−0.07)	37.82 (−0.01)	37.83 (−0.00)	**37.83**
Stockholm	37.84 (−0.12)	37.95 (−0.01)	37.96 (−0.00)	**37.96**
Park	39.51 (−0.59)	39.51 (−0.58)	39.68 (−0.41)	**40.09**
Parkrun	31.79 (−0.23)	31.99 (−0.03)	32.00 (−0.03)	**32.03**
Station2	43.61 (−1.33)	44.87 (−0.08)	44.94 (−0.00)	**44.94**
Mobcal	37.81 (−0.92)	38.64 (−0.08)	38.69 (−0.04)	**38.73**
Terrace	37.62 (−0.31)	37.93 (−0.00)	37.93 (−0.00)	**37.93**
Average	37.09 (−0.51)	37.42 (−0.18)	37.49 (−0.11)	**37.60**

[a]Motion fields estimated using MDP [3]

Table 6.1 shows the average per-sequence Y-PSNR obtained for FOHA-TFI without texture optimization (FOHA-TFI), with SWCA enabled (FOHA-TFI+W), as well as with both SWCA *and* optical blur synthesis enabled (FOHA-TFI+WO); as an anchor point, we further replicate the results obtained by BOA-TFI, which we presented in Sect. 4.4.3.

As mentioned earlier, the texture optimizations were designed primarily to improve the visual quality of the interpolated frames. As the results in the table show, the texture optimizations also improve in terms of Y-PSNR. It is worth highlighting that in all tested sequences, the results monotonously improve between FOHA-TFI without texture optimizations, FOHA-TFI+W, and FOHA-TFI+WO.

To conclude this section, we highlight the fact that the proposed motion-centric approach to frame interpolation allows us to readily identify regions that can benefit from selective smoothing of the predicted texture. This is in stark contrast to block-based TFI methods, which all have some inherent averaging applied to every pixel location of the upsampled frame.

6.4 Evaluation of TFI Performance

In this section, we provide a thorough evaluation of the performance of FOHA-TFI. Section 6.4.1 evaluates the method on a variety of common natural test sequences, and compares the scheme to three state-of-the-art TFI methods, as well as BOA-TFI. In Sect. 6.4.2, we "stress-test" the proposed method on highly challenging synthetic sequences from the Sintel dataset (see Sect. A.2), where the ground truth motion is known. We conclude the section with a comparison of the timings of the different TFI methods tested, which challenges some of the current "preconceptions" about the use of optical flow for TFI schemes.

6.4.1 Evaluation on Natural Sequences

In this section, we evaluate FOHA-TFI on a variety of common natural test sequences (see Sect. A.3), and compare the scheme both qualitatively *and* quantitatively to three state-of-the-art TFI methods, as well as BOA-TFI.

6.4.1.1 Quantitative Results

In this section, we evaluate the TFI performance of FOHA-TFI in terms of Y-PSNR of the interpolated frames. To confirm the statistical significance of the results, we conducted a *paired-samples t-test* to compare the Y-PSNR values of the proposed FOHA-TFI+WO method with each of the four other TFI methods. There was a *significant* difference in the scores for FOHA-TFI+WO and Jeong et al. [5] ($t(119) = 9.77$), Veselov and Gilmutdinov [6] ($t(119) = 14.87$), Lu et al. [7] ($t(119) = 9.58$), and BOA-TFI [4] ($t(119) = 11.66$); in all cases, $p < 0.001$, meaning high significance of the results. Table 6.2 shows the Y-PSNR values of the proposed method along with the three state-of-the-art TFI schemes. In the table, results that were statistically non-significant at an α value of 0.01 are highlighted. With respect to BOA-TFI (see Table 6.1), the results were significant for all but the "Shields1" and "Shields2" sequences.

To provide further insight into the quantitative results, Figs. 6.14 and 6.15 show plots of the Y-PSNR values on all tested frames individually for each of the 12 sequences. One can see that on most sequences, the proposed method constantly outperforms the others. The only statistically significant result where FOHA-TFI performs worse than existing state-of-the-art is Lu et al.'s [7] result on the "Kimono1" sequence. In this sequence, we observe that the motion boundaries between adjacent motion fields – estimated using MDP-flow [3] – do not always align, which troubles the motion-discontinuity based method for resolving double mappings. This highlights the importance of tailoring motion estimation schemes to the proposed

Table 6.2 Quantitative comparison of FOHA-TFI+WO with [5–7], on common natural test sequences. In parentheses (·), we show the difference between the Y-PSNR of the proposed FOHA-TFI+WO method and the respective method we compare it to ("−" means that the proposed FOHA-TFI+WO performs better, "+" means worse performance). Results where the paired-samples t-test between FOHA-TFI+WO and the respective method yielded non-significant *p-values* (at a level of $\alpha = 0.01$) are highlighted. **Bold** indicates per-row best performance

Sequence	Jeong [5]	Veselov [6]	Lu [7]	FOHA-TFI[a] +WO
Cactus	33.15 (−1.17)	31.27 (−3.04)	34.12 (−0.19)	**34.31**
Kimono1	33.93 (−0.09)	33.40 (−0.63)	**34.51** (+0.48)	34.03
Kimono2	39.97 (−1.67)	40.21 (−1.42)	39.51 (−2.12)	**41.63**
Rushhour	35.18 (+0.06)	34.93 (−0.19)	**35.30** (+0.19)	35.12
Shields1	35.90 (−0.68)	35.10 (−1.48)	35.89 (−0.69)	**36.58**
Shields2	33.87 (−3.96)	35.58 (−2.24)	33.52 (−4.31)	**37.83**
Stockholm	36.59 (−1.37)	37.12 (−0.84)	35.85 (−2.11)	**37.96**
Park	38.29 (−1.80)	38.84 (−1.26)	38.74 (−1.36)	**40.09**
Parkrun	30.63 (−1.40)	30.97 (−1.06)	30.50 (−1.53)	**32.03**
Station2	41.10 (−3.84)	41.41 (−3.54)	40.54 (−4.41)	**44.94**
Mobcal	29.13 (−9.60)	34.75 (−3.98)	29.53 (−9.20)	**38.73**
Terrace	33.29 (−4.64)	34.22 (−3.71)	33.66 (−4.26)	**37.93**
Average	35.08 (−2.51)	35.65 (−1.95)	35.14 (−2.46)	**37.60**

[a]Motion fields estimated using MDP [3]

scheme, which should be able to further improve the performance of the proposed method.

6.4.1.2 Qualitative Evaluation

While Y-PSNR comparisons are useful to summarize results, it is important to also look at the visual quality of the interpolated frames; for this reason, Figs. 6.16 and 6.17 take a closer look at two sequences.

All TFI methods investigated produce good results; however, FOHA-TFI in general shows superior performance. We observe that most comparisons in the literature are performed on sequences at CIF (352×288) resolution; furthermore, most of the standard test sequences are to varying degrees affected by motion blur. The sequences we use were recorded using high-quality, high-framerate cameras, and hence the individual frames contain much more high-frequency content. Block-based methods usually employ a variant of OBMC, which tends to oversmooth the interpolated frames, resulting in significant blurring of the overall texture; in Fig. 6.16, this can be seen in highly textured regions such as the text on the card of the Cactus sequence in the second row. Lu et al.'s [7] method seems to apply a particularly aggressive low-pass filter, which can explain the significant drop in Y-PSNR in some of the sequences.

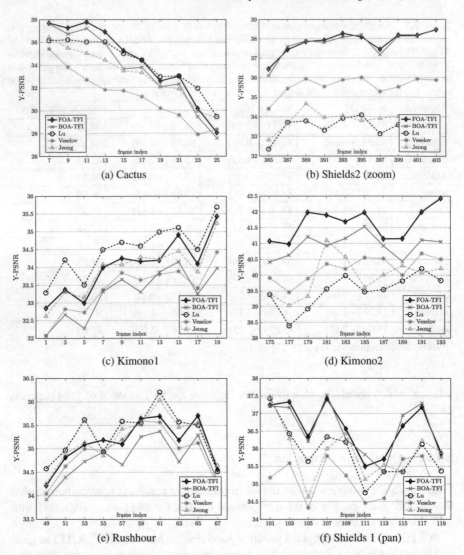

Fig. 6.14 Quantitative comparison of interpolated frame quality for the first half of the natural test sequences. Plots show the Y-PSNR of each individual interpolated frame for the proposed FOHA-TFI method, as well as Jeong et al. [5], Veselov and Gilmutdinov [6], Lu et al. [7], and our previously proposed BOA-TFI scheme [4]

Another important factor relates to the handling of regions around moving objects, as highlighted in the estimated disocclusion masks in Figs. 6.16c and 6.17c; such regions are only visible from one reference frame, and hence should only be predicted from the frame where they are visible. Besides our TFI methods, only [7] explicitly handles *occluded* regions. The quality of the proposed occlusion handling can be

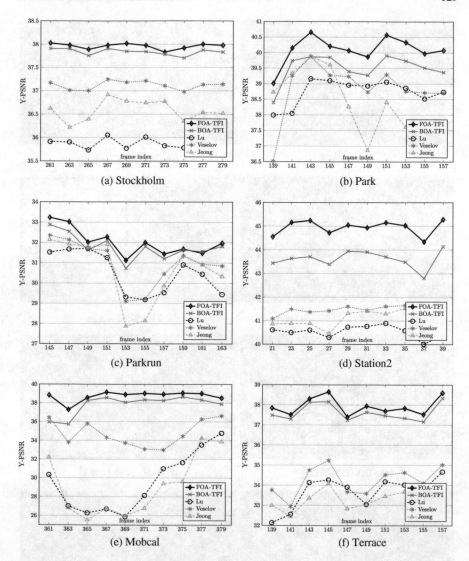

Fig. 6.15 Quantitative comparison of interpolated frame quality for the second half of the natural test sequences. Plots show the Y-PSNR of each individual interpolated frame for the proposed FOHA-TFI method, as well as Jeong et al. [5], Veselov and Gilmutdinov [6], Lu et al. [7], and our previously proposed BOA-TFI scheme [4]

appreciated in a number of sequences, but is most visible in the "Park" sequence in between the two trees, as well as the "Cactus" sequence, where the "Q" is properly interpolated by our method; both [5, 6] contain double edges, and [7] severely over-smooths this region (see Fig. 6.16h).

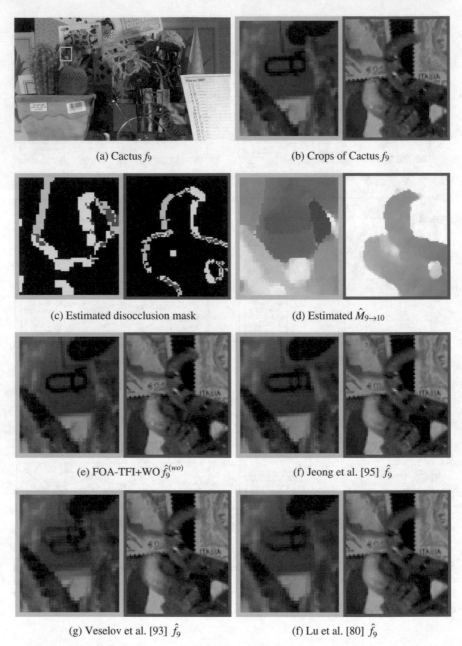

(a) Cactus f_9

(b) Crops of Cactus f_9

(c) Estimated disocclusion mask

(d) Estimated $\hat{M}_{9 \to 10}$

(e) FOA-TFI+WO $\hat{f}_9^{(wo)}$

(f) Jeong et al. [95] $\hat{f}_9$

(g) Veselov et al. [93] $\hat{f}_9$

(f) Lu et al. [80] $\hat{f}_9$

Fig. 6.16 TFI comparison on "Cactus" sequence. The first shows the ground truth frame (**a**), as well as crops of the ground truth frame (**b**); **c** shows the estimated forward and reverse disocclusion masks (cyan and yellow, respectively), and **d** shows the (colour-coded) mapped forward motion field $\hat{M}_{b \to c}$. **e–h** show crops of the proposed FOHA-TFI+WO (with both texture optimizations), Jeong et al. [5], Veselov et al. [6], and Lu et al. [7], respectively

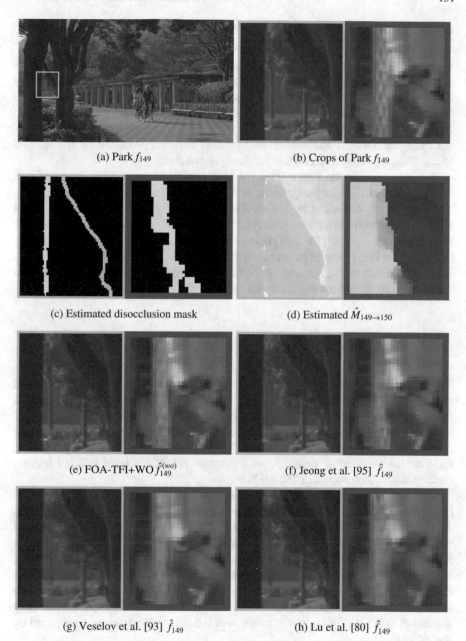

(a) Park f_{149}

(b) Crops of Park f_{149}

(c) Estimated disocclusion mask

(d) Estimated $\hat{M}_{149 \to 150}$

(e) FOA-TFI+WO $\hat{f}_{149}^{(wo)}$

(f) Jeong et al. [95] $\hat{f}_{149}$

(g) Veselov et al. [93] $\hat{f}_{149}$

(h) Lu et al. [80] $\hat{f}_{149}$

Fig. 6.17 TFI comparison on "Park" sequence. The first shows the ground truth frame (**a**), as well as crops of the ground truth frame (**b**); **c** shows the estimated forward and reverse disocclusion masks (yellow and cyan, respectively), and **d** shows the (colour-coded) mapped forward motion field $\hat{M}_{b \to c}$. **e–h** show crops of the proposed FOHA-TFI+WO (with both texture optimizations), Jeong et al. [5], Veselov et al. [6], and Lu et al. [7], respectively

6.4.2 Evaluation on Synthetic Sequences

In the previous sections, we have evaluated the performance of FOHA-TFI with *estimated* motion fields from a state-of-the-art optical flow method. Since the quality of these motion fields is out of our control, we find it useful to provide additional results on a challenging set of computer-generated sequences, where the ground truth motion is known. Figure 6.18 shows the forward inferred motion field ($\hat{M}_{b\to c}$), the *estimated* forward and reverse disocclusion mask as produced by our method ($\hat{S}_{b\to a} \cup \hat{S}_{b\to c}$), as well as crops of the interpolated frame $\hat{f}_b$.

The even rows in the figure show the high quality of the *warped* motion fields for these very challenging sequences, in particular around moving objects; the disocclusion masks *estimated* by FOHA-TFI expose the amount of disocclusion that arises between two consecutive frames of the various sequences. In the odd rows, we show crops of the motion fields in challenging parts, together with the corresponding regions of the interpolated frames produced by FOHA-TFI. The high-quality results are a culmination of high quality warped motion fields, high-precision disocclusion masks, as well as the two texture optimizations.

6.4.3 Processing Times

In this section, we report on the processing times of FOHA-TFI, and compare it with [5–7], as well as BOA-TFI. We note that none of the methods are optimized for time, and, with the exception of BOA-TFI, the timings were obtained on different machines, kindly provided by the authors of the respective TFI method. The relevant specifications of the testing machines, as well as the average per-frame processing time, are summarized in Table 6.3; we further provide the average PSNR for every method tested, as reported in Table 6.2.

We split up the processing times for the *motion estimation (ME)* part and the *frame interpolation (FI)* part, and note that our contribution lies solely in the FI part. As mentioned before, both FOHA-TFI *and* BOA-TFI can be used with any optical flow method (ME part) that produces sharp discontinuities around moving objects; in this thesis, we primarily used *motion detail preserving (MDP)* optical flow [3] to estimate motion fields, which produces very high-quality optical flow fields, at the expense of comparatively high processing times. To give some more insight into the impact of different optical flow estimators, we further ran the experiments using Revaud et al.'s [8] EPIC flow algorithm, using the default parameters proposed by the authors.

One can observe that EPIC flow [8] runs about 50 times faster than MDP [3], and 4 times faster than the fastest competitor [6], while still outperforming current state-of-the-art methods by more than 1dB. As these results show, block-based TFI

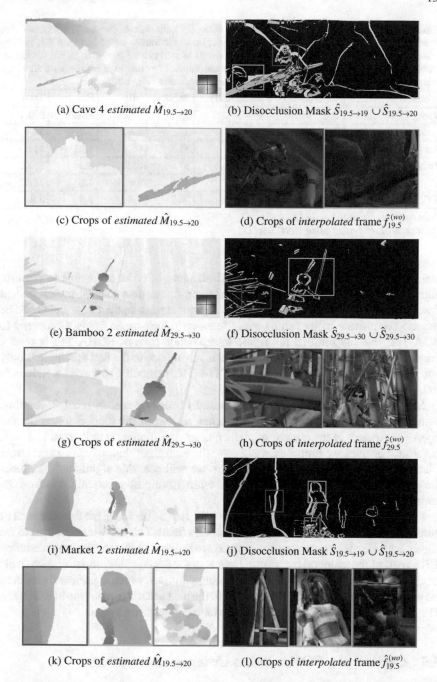

(a) Cave 4 *estimated* $\hat{M}_{19.5 \to 20}$

(b) Disocclusion Mask $\hat{S}_{19.5 \to 19} \cup \hat{S}_{19.5 \to 20}$

(c) Crops of *estimated* $\hat{M}_{19.5 \to 20}$

(d) Crops of *interpolated* frame $\hat{f}_{19.5}^{(wo)}$

(e) Bamboo 2 *estimated* $\hat{M}_{29.5 \to 30}$

(f) Disocclusion Mask $\hat{S}_{29.5 \to 30} \cup \hat{S}_{29.5 \to 30}$

(g) Crops of *estimated* $\hat{M}_{29.5 \to 30}$

(h) Crops of *interpolated* frame $\hat{f}_{29.5}^{(wo)}$

(i) Market 2 *estimated* $\hat{M}_{19.5 \to 20}$

(j) Disocclusion Mask $\hat{S}_{19.5 \to 19} \cup \hat{S}_{19.5 \to 20}$

(k) Crops of *estimated* $\hat{M}_{19.5 \to 20}$

(l) Crops of *interpolated* frame $\hat{f}_{19.5}^{(wo)}$

Fig 6.18 TFI results on challenging synthetic sequences from the Sintel dataset. **a, e,** *i*) show the estimated forward motion fields produced by our method; **b, f, j**) show the union of the *estimated* forward and reverse disocclusion masks, where yellow means not visible in f_a, cyan means not visible in f_c, and red are regions that are not visible in either reference frame. **c, g, k** show crops of the estimated motion fields, and **d, h, l** the interpolated target frames produced by FOHA-TFI+WO

Table 6.3 Comparison of processing times of FOA-TFI with state-of-the-art TFI schemes. We show the average per-frame processing time (in sec) on all the frames tested in Sect. 6.4.1, split up in *motion estimation (ME)* and *frame interpolation (FI)*, as well as total processing time. We further provide the CPU and amount of RAM of the machines used to obtain the results, as well as the average Y-PSNR obtained on the whole test set. **Bold** indicates best performance

Method	CPU (GHz)	RAM (GB)	PSNR	ME	FI	Total
Jeong [5]	2.8	8	35.1	410.2	498.9	909.1
Veselov [6]	2.6	8	35.7	32.4	2.1	34.5
Lu [7]	2.4	6	35.1	96.2	18.1	114.3
BOA-TFI [4][a]	3.2	8	37.1	355.4	8.2	363.6
FOHA-TFI+WO[a]	3.2	8	**37.6**	355.4	**1.8**	357.2
FOHA-TFI+WO[b]	3.2	8	36.7	**7.0**	**1.8**	**8.8**

[a]Motion fields estimated using MDP [3]
[b]Motion fields estimated using EPIC flow [8]

methods need a lot of constraints and optimizations in order to be able to perform high-quality TFI, which in the end turn out to be slower than state-of-the-art optical flow methods. Furthermore, as manifested by our results, optical flow fields can be expected to provide better results than "fixed-up" block-based fields. Compared to BOA-TFI, one can see that the frame interpolation time of FOHA-TFI is over 4 times shorter than that of BOA-TFI, which is mostly due to the fact that there is only one motion field inversion involved in FOHA-TFI, compared to three inversions in BOA-TFI.

Motion estimation left aside, the proposed FI method spends most of the time mapping triangles from one frame to another, as part of the motion *inversion* and *inference* process. In the current implementation, the triangle size is fixed to 1×1. In the next chapter, we propose a mesh sparsification algorithm, which creates larger triangles in regions of smooth motion. As we will see, this significantly reduces the processing time of the TFI procedure, while having no practical impact on the interpolation quality.

In existing video codecs, the motion has to be (re-)estimated at the decoder for TFI purposes, which constrains the quality of motion fields. This is in stark contrast to the highly scalable video compression systems we propose in this thesis, which employ TFI as part of the temporal transform. Hence, at the decoder, high-quality "physical" motion can be employed for frame upsampling purposes, which significantly reduces the processing time of the TFI framework. In the next section, we outline how FOHA-TFI can be extended to a highly scalable video compression system.

6.5 Outline of a FOHA Video Compression System

The focus of this chapter was on three modifications of the BOA-TFI scheme to further improve the quality of the interpolated frames. We have performed an extensive comparison of the TFI performance of both FOHA-TFI and BOA-TFI, and shown

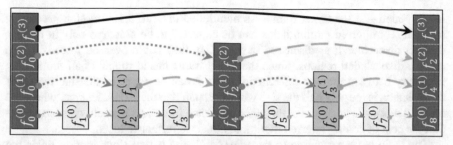

(a) Forward-only hierarchical anchoring of motion (FOHA).

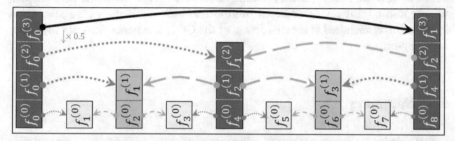

(b) Bidirectional hierarchical anchoring of motion (BIHA).

Fig. 6.19 Comparison of the way the *coded* motion fields are "anchored" in **a** FOHA and **b** BIHA

how the proposed changes positively impact the interpolation quality. At this point, a logical next step would be to incorporate the FOHA-TFI scheme into a highly scalable video compression scheme, similar to what we have done in Chap. 5, where BOA-TFI formed the essential building block of the BIHA scheme; we call this scheme *forward-only hierarchical anchoring (FOHA)*.

Since the incorporation requires quite some effort from an implementation point of view, we decided not to pursue this undoubtedly promising direction. In this section, instead, we give an outline of what a video compression system based on FOHA-TFI could look like, and briefly discuss anticipated advantages over the BIHA scheme. We also highlight problems that would be present in the FOHA scheme, which lead us to the proposal of a third motion anchoring strategy, which we present in the next chapter.

Figure 6.19 shows the FOHA scheme; for comparison, we also show the BIHA scheme. As can be seen in the figure, in the FOHA scheme, as its name suggests, all *coded* motion fields are pointing forward. We have already extensively discussed the advantages of FOHA-TFI over BOA-TFI, namely guaranteed temporal consistency of the warped motion fields, more robust disambiguation of disoccluded regions and improved disocclusion region handling. Furthermore, the proposed texture optimizations can also be expected to reduce the prediction residual and hence improve the compression performance of the FOHA scheme. One aspect that is particular to the compression scenario is the fact that the *inferred* motion fields are anchored – and hence coded – at the target frame. As mentioned in Sect. 5.1.2, motion prediction residuals for inferred motion fields can be expected to be non-zero only in regions that get disoccluded; as discussed in Sect. 5.6, they could even be forced to zero outside disoccluded regions. Since the target frame lies at roughly half the motion trajectory between the two reference frames, it can be expected that the region of disocclusion is roughly half the size when compared to the BIHA scheme, where the inferred motion is anchored at the other reference frame. As a result, one can expect that the cost of coding inferred motion residuals is further reduced.

The main issue pertaining to the FOHA scheme is that since motion fields are coded at different frames, small rounding errors are inevitable. In the next chapter, we propose a third reference-based motion anchoring strategy, where all motion information is anchored at the first frame of the GOP, and hence does not introduce rounding-errors.

6.6   Summary

In this chapter, we proposed three modifications to the BOA-TFI scheme presented in Chap. 4. First, we proposed a more robust measure of motion discontinuity based on the divergence of motion fields. The "so-called" DFLM allows us to handle regions of complex geometry better than with the *binary* motion discontinuity representation induced from breakpoints. Second, we flipped the anchoring of the inferred motion fields; we coined the resulting TFI scheme FOHA-TFI. A main advantage of FOHA-TFI over BOA-TFI is that it needs only one motion field inversion, as opposed to three motion inversions in the case of BOA-TFI. This reduces the computational complexity by a factor of 3. At the same time, it *guarantees* that the warped forward and backward pointing prediction fields are *geometrically consistent*, which further improves the prediction quality of the proposed scheme. Lastly, we proposed two effective texture optimizations, which selectively smooth regions where the scheme can be expected to create artificial high frequencies. A comprehensive experimental evaluation of the FOHA-TFI scheme on both natural and synthetic sequences showed the high performance of the proposed TFI framework compared to state-of-the-art TFI methods.

While the focus of this chapter was on frame interpolation, which is the essential building block of the proposed highly scalable video compression scheme, we outlined a potential scalable video coder, which we referred to as FOHA. In the next chapter, we present a third, further simplified reference-based motion anchoring strategy, which has a number of highly interesting properties for (scalable) video compression.

References

1. J.-R. Ohm, *Multimedia Communication Technology: Representation, Transmission and Identification of Multimedia Signals* (Springer Science & Business Media, Berlin, 2012) (cited on pages 108, 109)
2. D.J. Butler, J. Wulff, G.B. Stanley, M.J. Black, *A Naturalistic Open Source Movie for Optical Flow Evaluation*, in *European Conference on Computer Vision* (2012), pp. 611–625 (cited on page 127)
3. L. Xu, J. Jia, Y. Matsushita, Motion detail preserving optical flow estimation. IEEE Trans. Patt. Anal. Mach. Intell. 1744–1757 (2012). (cited on pages 43, 44, 49, 73, 95, 102, 128, 130, 131, 134, 137, 139, 159, 185–187)
4. L. Xu, J. Jia, Y. Matsushita, Occlusion-aware temporal frame interpolation in a highly scalable video coding setting, in *APSIPA Transactions on Signal and Information Processing*, vol. 5 (2016) (cited on pages vii, 54, 67, 69, 74, 75, 130, 132, 133, 139)
5. S.-G. Jeong, C. Lee, and C.-S. Kim, Motion-Compensated Frame Interpolation based on multihypothesis motion estimation and texture optimization, in *IEEE Transactions on Image Processing* (2013), pp. 4497–4509. (cited on pages 47, 49, 73–75, 131–137, 139)
6. A. Veselov, M. Gilmutdinov, Iterative hierarchical true motion estimation for temporal frame interpolation, in *IEEE International Workshop on Multimedia Signal Processing* (2014) (cited on pages 47, 49, 73–75, 131–137, 139)
7. Q. Lu, N. Xu, X. Fang, Motion-compensated frame interpolation with multiframe based occlusion handling. IEEE J. Disp. Tech., **11**(4) (2015). (cited on pages 41, 42, 48, 49, 73-76, 131–137, 139)
8. J. Revaud, P. Weinzaepfel, Z. Harchaoui, C. Schmid, EpicFlow: edge-preserving interpolation of correspondences for optical flow, in *Proceedings IEEE Conference on Computer Vision and Pattern Recognition* (2015). (cited on pages 43, 44, 137, 139)

Chapter 7
Base-Anchored Motion (BAM)

In the previous two chapters, we have presented the BIHA and the FOHA scheme for highly scalable video compression. Both these schemes employ reference-based *hierarchical* motion field anchorings; that is, at each level of the temporal hierarchy, motion fields are *predictively* coded. For this, motion fields from coarse levels are mapped to the same or finer temporal levels in order to predict motion information. We have seen that in FOHA, the geometrical consistency of the mapped motion fields can be guaranteed. However, there is no way of ensuring that multiple interpolated frames are interpolated in a *temporally consistent* way. Furthermore, as observed at the end of Sect. 6.5, the problem is that the mapped motion information in general does not fall onto integer grid locations, and hence rounding errors are inevitable.

In this chapter, we explore a *base-anchored motion (BAM)* scheme,[1] which addresses the aforementioned issues. In this motion anchoring strategy, all *coded* motion information is anchored at one *base frame*, which is the first frame of the GOP. Figure 7.1 shows the BAM scheme, together with the other two motion anchoring strategies proposed in this thesis. In the previous anchoring strategies, the CAW procedure we use to map motion information from one frame to another uses a fixed cell size of 1×1; as we mentioned earlier, the cell size could be increased in regions of smooth motion. However, with hierarchically anchored motion as employed in BIHA and FOHA, such motion sparsification has to be performed at various frames, which is both computationally less efficient, and, perhaps more importantly, damages the geometrical (and temporal) consistencies we try to enforce. In the BAM framework, motion sparsification makes more sense, since all motion information is described in the same grid; we present a simple *mesh sparsification* algorithm in Sect. 7.2.

---

[1]The work introducing the BAM scheme is accepted for publication at the IEEE International Workshop on Multimedia Signal Processing (MMSP) [1].

© Springer Nature Singapore Pte Ltd. 2018
D. Rüfenacht, *Novel Motion Anchoring Strategies for Wavelet-Based Highly Scalable Video Compression*, Springer Theses,
https://doi.org/10.1007/978-981-10-8225-2_7

139

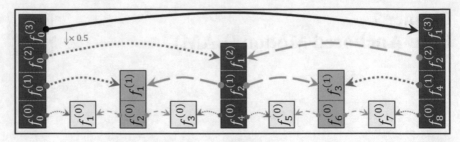

(a) Bidirectional hierarchical anchoring of motion (BIHA).

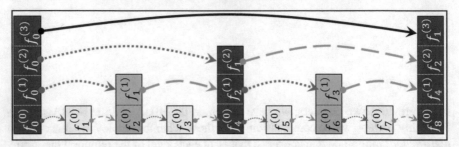

(b) Forward-only hierarchical anchoring of motion (FOHA).

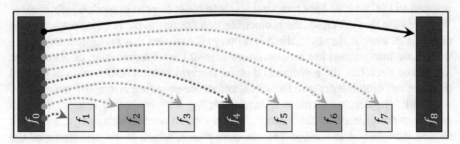

(c) Base-Anchored Motion (BAM).

Fig. 7.1 Comparison of the three motion field anchoring strategies investigated in this thesis. As before, arrows indicate motion fields that are *coded*. One can see how in BAM, all coded motion fields are anchored at the first frame of the GOP, which enables a very compact representation of the motion

In the BAM scheme, the base frame holds motion information linking it with *any* other frame of the GOP, as depicted in Fig. 7.1c. Combined with the motion inference operations presented in earlier chapters, *temporally consistent* motion information can be readily composed between *any* two frames of the GOP.[2] Another advantage of the centralized motion organization is that motion information can be more compactly represented. In Sect. 7.3, we explore the compact motion representation by adding

---

[2]The evaluation of temporal consistency in this thesis is limited to qualitative results, since there exists no good quantitative assessment tool.

higher-order motion models to the framework, which are able to better describe the "true" trajectory of objects through the spatio-temporal volume.

It is important to highlight the fact that while the *motion anchoring* is no longer hierarchical, the *temporal transform* can still be performed in a hierarchical way. That is, texture information other than the one from the coarsest temporal level can be used to predict the target frames at finer temporal levels. For the initial exploration of BAM in this thesis, however, we use a "flat" prediction structure, as depicted in Fig. 7.1c; that is, any target frame of the GOP is predicted from only the two coarsest level frames (i.e., the first and the last frame of the GOP). In Sect. 7.4, we show how the BAM scheme with the flat prediction structure can be integrated into HEVC. That is, we use BAM to perform the motion-compensation, and use the highly optimized compression framework offered by HEVC to code all the texture residuals. In the following, we use the example of TFI to show how the BAM scheme works.

7.1 High Framerate Upsampling with BAM

In this section, we present the BAM framework using the application of high framerate upsampling. We start with an overview of the scheme, which is guided by Fig. 7.2, and then present the fundamental changes with respect to the FOA-TFI scheme. Input to the method are two reference frames f_0 and f_1, where we *normalized* the time interval for ease of notation. Furthermore, the scheme requires *at least* the motion field describing the trajectory from f_0 to f_1, denoted as $M_{0 \to 1}$.[3] For the remainder of this chapter, we use $\mathbf{u}_{0 \to t}(\mathbf{m})$ to denote a motion vector which links a location $\mathbf{m}$ in f_0 with its corresponding location in f_t. As we shall see in Sect. 7.3, the BAM framework facilitates the incorporation of higher order motion models. We hence further use $\mathbf{u}_{0 \to t}^{(n)}(\mathbf{m})$ to denote a motion vector following an nth-order motion model, and refer to the framework that employs nth-order motion as $\text{BAM}^{(n)}$. For ease of notation, we drop the model order superscript whenever the distinction between model orders is not necessary.

From the two reference frames and a motion description between the two, the aim is to interpolate frames f_t at the *normalized time instance* $t \in [0, 1]$ in between the two reference frames f_0 and f_1; for example, the standard case of doubling the framerate that is commonly considered in the literature is obtained by setting $t = 0.5$. We highlight that the proposed framework can be used for arbitrary upsampling factors, where the centralized motion organization allows us to make *temporally consistent decisions* for all interpolated frames.

In the first step, the base motion field $M_{0 \to 1}$ is partitioned into a *base mesh* $\mathcal{B}_0$, which holds a collection of K vertices $\{V_0^k\}, k \in \{0, K - 1\}$, each of which holds

---

[3]As in the methods presented in the previous chapters, the quality of the proposed TFI scheme depends on the quality of the input motion fields. In particular, the motion fields need to have sharp discontinuities around moving object boundaries; this is because we use motion discontinuity information to reason about foreground objects in order to resolve double mappings and disoccluded regions.

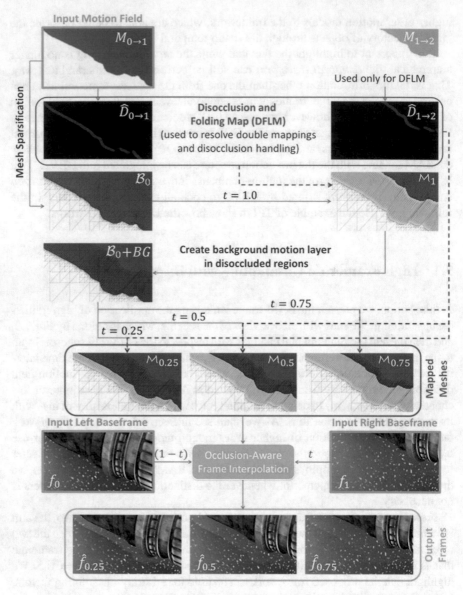

Fig. 7.2 Overview of the proposed BAM framework. Input to the method are two reference frames f_0, and f_1, and a motion field $M_{0\rightarrow1}$ between the two reference frames. We propose a triangular mesh sparsification algorithm, which reduces the computational complexity of the subsequent motion mapping operations. The base mesh $\mathcal{B}_0$ is mapped to the right reference frame, where all disoccluded regions are observed. These are then mapped back to the left reference frame, where they form a background motion layer. Mapping the mesh to any intermediate target frame f_t creates temporally consistent motion in disoccluded regions. Using reasoning about motion boundaries, we resolve regions in f_t where multiple triangles overlap. In the last step, f_t is bidirectionally predicted from both reference frames; regions that are only visible in one reference frame are detected and unidirectionally predicted

motion vectors $\{\mathbf{u}_{0\rightarrow t}(V_0^k)\}$ "linking" f_0 with *any* f_t we wish to interpolate (under constant motion assumption). In addition, the vertices hold motion vectors for the special case of $t = 1$, which links f_0 with f_1. These vertices are connected to form triangles, whose affine motion approximates the underlying motion field. In the extreme case, each integer location of the motion field gets assigned a vertex, which results in triangles of size 1×1; this is essentially what happened in the CAW procedure presented in Sect. 4.2. However, while around moving object boundaries, even a triangle of size 1×1 is not able to describe the motion because of the discontinuity, one can expect that the affine motion from relatively large triangles is able to well approximate the smooth motion within objects. In Sect. 7.2, we present a *mesh sparsification* algorithm, which creates a mesh with variable triangle sizes.

In the proposed scheme, we also use the concept of a *mapped* mesh $\mathcal{M}_t$, which is obtained by mapping the base mesh $\mathcal{B}_0$ to frame f_t. The main *difference* between a mapped mesh and the base mesh are the motion vectors their vertices hold: the mapped mesh contains vertices whose motion vectors $\mathbf{u}_{t\rightarrow 0}(V_t^k)$ and $\mathbf{u}_{t\rightarrow 1}(V_t^k)$ link it with both the preceding *and* the succeeding reference frames f_0 and f_1, respectively; this is illustrated in Fig. 7.3. We come back to how motion vectors in mapped meshes are assigned in Sect. 7.1.1. An important difference to the anchoring schemes presented in the earlier chapters is that the base anchoring allows us to establish links with triangles; that is, each triangle of the mesh is associated with a *triangle identifier* *(TID)*, which makes it possible to keep track of where the triangles map through the spatio-temporal volume. We will show in Sect. 7.1.4 how this information can be used to create *visibility masks* that inform which triangles are visible in each of the reference frames.

Setting $t = 1$, we map the base mesh from f_0 to f_1; the importance of the obtained $\mathcal{M}_1$ is that it reveals regions that get disoccluded between the two reference frames. Those are the regions for which affine interpolation results in a "non-physical" motion vector assignment. In the case of frame upsampling factors larger than two (i.e., more than one frame is interpolated between any pair of reference frames), it is particularly important that *temporally consistent* motion is assigned in such regions. In the proposed scheme, this is achieved by mapping the disoccluded regions back

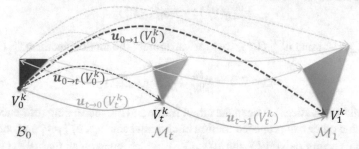

Fig. 7.3 Illustration how triangles of the base mesh and mapped meshes are linked via motion vectors. Dashed red arrows show base anchored motion, whereas solid green arrows show mapped mesh motion for $\mathcal{M}_t$, which links each vertex of the mesh with the preceding and succeeding reference frame

to the base mesh $\mathcal{B}_0$, which essentially creates *local background motion layers*. The purpose of these local background layers is that if mapped to intermediate frames f_t in between the two base frames, temporally consistent motion will be assigned in disoccluded regions for varying values of t; this procedure is explained in more detail in Sect. 7.1.2.

In the mapped mesh $\mathcal{M}_t$, triangles overlap in regions where a foreground object moves on top of a background object. Like in FOA-TFI, we use the DFLM (second row of Fig. 7.2) to resolve such double mappings. Using a *triangle identifier (TID)* compatibility check to identify regions that are not visible in either of the reference frames, we are able to switch from bidirectional to unidirectional prediction of the interpolated target frame(s) f_t in regions that are only visible in one of the reference frames. In the following, we explain the main changes to the framework presented in the previous chapter in more detail.

7.1.1 Affine Mesh Warping

We now describe how the base mesh $\mathcal{B}_0$ is mapped to any *normalized* time instance $t \in]0, 1]$ in between the two reference frames f_0 and f_1, where it will form the mapped mesh $\mathcal{M}_t$. For the description of the mapping process, we focus on how an individual vertex V_0^k is mapped to the target frame; the mapped vertices can then be connected together to form triangles, which completely cover the target frame. Motion $\hat{M}_{t \to 0}^{(n)}$ and $\hat{M}_{t \to 1}^{(n)}$, which relates f_t with its preceding and succeeding reference frame f_0 and f_1, respectively, can then be obtained through affine interpolation of the motion vectors $\mathbf{u}_{t \to 0}^{(n)}(V_t^k)$ and $\mathbf{u}_{t \to 1}^{(n)}(V_t^k)$, respectively.

We use $\mathbf{u}_{0 \to t}^{(n)}(V_0^k)$, derived from an nth-order motion model as described in Sect. 7.3, to map V_0^k to the target frame; that is,

$$V_t^k = V_0^k + \mathbf{u}_{0 \to t}^{(n)}(V_0^k). \tag{7.1}$$

Negating its motion yields the motion linking the target frame with the preceding reference frame f_0:

$$\mathbf{u}_{t \to 0}^{(n)}(V_t^k) = -\mathbf{u}_{0 \to t}^{(n)}(V_0^k). \tag{7.2}$$

From $\mathbf{u}_{0 \to 1}^{(n)}(V_0^k)$ and $\mathbf{u}_{0 \to t}^{(n)}(V_0^k)$, we "forward infer" the motion vector $\mathbf{u}_{t \to 1}^{(n)}(V_t^k)$:

$$\mathbf{u}_{t \to 1}^{(n)}(V_t^k) = \mathbf{u}_{0 \to 1}^{(n)}(V_0^k) - \mathbf{u}_{0 \to t}^{(n)}(V_0^k), \tag{7.3}$$

which links the vertices between the target frame and the succeeding reference frame f_1. As in the other two proposed motion anchoring schemes, $\mathbf{u}_{t \to 1}^{(n)}(V_t^k)$ is composed of motion vectors $\mathbf{u}_{0 \to 1}(V_0^k)$ and $\mathbf{u}_{t \to 0}^{(n)}(V_t^k)$, which guarantees that $\mathbf{u}_{t \to 0}^{(n)}(V_t^k)$ and $\mathbf{u}_{t \to 1}^{(n)}(V_t^k)$ point to the same geometrical location in f_0 and f_1. In addition, the central motion organization enables to make *temporally consistent* motion assignments between the set of interpolated target frames $\{f_t\}$.

One can identify the motion inversion (7.2) and forward inference (7.3) motion operations that are used in the FOA-TFI scheme. In contrast to FOA-TFI, where triangles were individually mapped and eventual ambiguities were resolved "on the spot", here, all vertices of the base mesh are mapped to form the mapped mesh $\mathcal{M}_t$, and only then regions of disocclusion and double mappings are handled. This opens up interesting optimization operations in the future, such as better (global) handling of disoccluded regions.

In the mapped mesh $\mathcal{M}_t$, on the *leading* side of moving objects, there will be regions where foreground triangles overlap with background triangles. Furthermore, in regions on the *trailing* side of objects in motion (disocclusion), the affine interpolated motion between foreground and background vertices is non-physical. In Sect. 7.1.2, we show how to assign more "physical" and temporally consistent motion in disoccluded regions; in Sect. 7.1.3, we explain how the foreground triangle can be identified in regions where multiple triangles overlap.

7.1.2 Temporally Consistent Motion in Disoccluded Regions

The higher the frame upsampling factor, the more critical a temporally consistent interpolation in disoccluded regions becomes, since inconsistent motion can lead to visually disturbing artefacts. Most existing TFI methods have no way of guaranteeing consistent interpolation in disoccluded regions.

Using Fig. 7.4, where a rectangle moves on top of static background, we now provide a high-level overview of the proposed procedure to assign temporally consistent motion in disoccluded regions. The method is most easily understood in the case of constant velocity motion, in which case the area of disocclusion monotonically increases as objects transition from the preceding reference frame f_0 to the succeeding reference frame f_1. This implies that mapping the motion from f_0 to f_1 exposes all regions that get disoccluded between these two frames, as shown in Fig. 7.4b (yellow area). The idea of the proposed method is to extrapolate (local) background motion in all disoccluded regions, and then map these regions back to the left base frame f_0, where they are added to the base mesh $\mathcal{B}_0$. In doing this, we effectively generate a (set of) local background motion layer(s), as illustrated in Fig. 7.4c.

During the affine mesh warping procedure (see previous section), the newly created triangles forming the background layer(s) are mapped in the same way as the "original" triangles of the sparsified mesh. In any region that is not hit by any other triangle, the local background layer will guarantee that temporally consistent motion is assigned, as illustrated in Fig. 7.4d.

We now provide more technical details how this background motion layer can be created. We traverse all triangles of the base mesh $\mathcal{B}_0$. For each triangle of size 1×1 (i.e., triangles that cross motion discontinuities), we identify the *edges* that traverse a large (positive) divergence, which is indicative of disocclusion; we call such edges *disocclusion split edges (DSE)*. In the following, we focus on one such DSE, noting

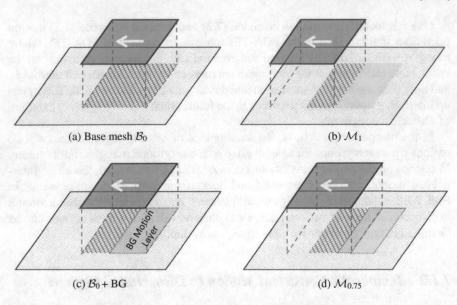

(a) Base mesh $\mathcal{B}_0$ (b) $\mathcal{M}_1$

(c) $\mathcal{B}_0$ + BG   (d) $\mathcal{M}_{0.75}$

Fig. 7.4 High-level illustration of the proposed disocclusion region motion back-propagation method. Mapping the base mesh $\mathcal{B}_0$ in to the next reference frame f_1 reveals all regions that get disoccluded between the two reference frames, indicated by the yellow area in **b**. Mapping these regions back to the left base frame creates a background motion layer, as shown in **c**, which guarantees temporally consistent motion assignments in disoccluded regions for *any* target frame f_t, e.g., $t = 0.75$ in **d**

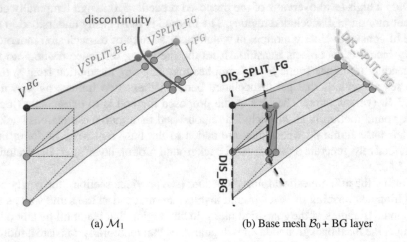

(a) $\mathcal{M}_1$ (b) Base mesh $\mathcal{B}_0$ + BG layer

Fig. 7.5 Illustration of how disoccluding triangles are split up at motion discontinuities. **a** on each split edge, two new vertices are created at the location of maximum divergence (discontinuity); one gets assigned the motion of the background vertex, which we label V^{SPLIT_BG}; similarly, V^{SPLIT_FG} gets assigned extrapolated motion from the foreground vertex. Mapping these vertices back to $\mathcal{B}_0$ and connecting them to form triangles creates a background motion layer

that the same procedure is applied to all DSEs. We illustrate the procedure using Fig. 7.5, which focuses on one disoccluding triangle (yellow).

First, the DSE is mapped to the right reference frame (Fig. 7.5a), where we use motion divergence information at frame f_1 to search for the location of maximum divergence; this is where we set the *split location*. Following the reasoning that motion discontinuities travel with the foreground object, we label the vertex of the edge that is closer to the point of maximum divergence (i.e., the motion discontinuity) as foreground V^{FG} (green circle), whereas the other one is labelled as background vertex V^{BG} (red circle). Next, we create two (disconnected) vertices at the split location; we label one of the two newly created "split vertices" as V^{SPLIT_BG} (orange circle), and the other one as V^{SPLIT_FG} (cyan circle). The motion assigned to these split vertices is obtained by extrapolating the motion of the corresponding V^{BG} and V^{FG} vertices in $\mathcal{M}_1$.

In the next step, the split vertices are mapped back to the left base mesh $\mathcal{B}_0$. After all the "split edges" are split up and the split vertices have been mapped back to $\mathcal{B}_0$, we form new triangles by connecting all V^{SPLIT_FG} and V^{SPLIT_BG} vertices with the corresponding V^{FG} and V^{BG}, respectively, as shown in Fig. 7.5b. Hypothetically connecting all *adjacent* V^{SPLIT_BG} together results in a delineation of the "end" of the disoccluded region, which we call "DIS_SPLIT_BG" in the figure (yellow dashed line); together with the "DIS_BG" line, they outline the local background motion layer.

The method described above is only valid for $BAM^{(1)}$ (i.e., constant velocity motion), since otherwise regions of disocclusion are not guaranteed to increase monotonically between the two base frames f_0 and f_1. Extensions to account for higher-order motion models are possible, but are out of the scope of our initial investigation presented in this thesis. As the experimental validation in Sect. 7.3.1 shows, even assuming constant velocity in disoccluded regions is able to provide good results for $BAM^{(2)}$.

7.1.3 Computing a Foreground Triangle ID Map

In the mapped mesh $\mathcal{M}_t$, triangles belonging to the foreground object can overlap with triangles of the background object; this happens on the leading side of moving objects. The reasoning to resolve such double mappings is the same as the one used in the FOA-TFI scheme, which we presented in Sect. 6.2.3; the main difference is that whereas the double mapping disambiguation in FOA-TFI was done on a per-pixel basis, in BAM it is done per triangle. That is, for each triangle that is overlapping with another triangle in the mapped mesh, we sample its affine motion to obtain motion hypothesis 1. A triangle potentially overlaps with multiple triangles; we sample the motion of one of the intersecting triangles to create motion hypothesis 2. The identification of the foreground motion hypothesis is identical to the one presented in Sect. 6.2.3. If the current triangle is found to be in the foreground, any integer location **m** that is covered by the identified foreground triangle gets assigned the corresponding triangle ID in a *foreground TID map*, denoted as F_t. In the next section, we show how TIDs can be used to determine which regions of the reference frames are visible in the target frame.

7.1.4   Assessing Visibility Using Triangle ID Compatibility Check

We use TIDs to assess whether a particular location **m** is hidden in either of the reference frames. This is done by a *TID compatibility check*, which tests whether the identified foreground TID at location **m** in f_t (i.e., $F_t[\mathbf{m}]$) is the same as the one if **m** is mapped to f_0 or f_1 using the affine interpolated motion of the specific triangle. We store this valuable information in *visibility masks* $I_t^p[\mathbf{m}]$, where $p = 0$ or $p = 1$ is used to distinguish between the reverse and forward visibility masks, respectively. More precisely,

$$I_t^p[\mathbf{m}] = \begin{cases} 1 & F_t[\mathbf{m}] = F_p\big[\mathbf{m} + \hat{M}_{t \to p}[\mathbf{m}]\big] \\ 0 & \text{otherwise} \end{cases}, \qquad (7.4)$$

where $\hat{M}_{t \to p}[\mathbf{m}]$ is the affine interpolated motion of the identified *foreground* triangle that covers location **m**.

We illustrate the creation of the forward and reverse visibility masks with the example shown in Fig. 7.6; in the figure, we use colours to differentiate between TIDs. The yellow triangle sits in a region of forward disocclusion, which means that it is not visible in the preceding reference frame f_0. Applying the (background) motion to a location **m** within the yellow triangle in f_t maps to a location where the foreground TID is different (orange triangle), and hence it is marked as not visible in the reverse visibility mask ($I_t^0[\mathbf{m}] = 0$); mapping the same location **m** forward, on the other hand, yields a positive visibility check, and hence $I_t^1[\mathbf{m}] = 1$. The same reasoning can be applied for reverse disocclusions (cyan region), where the TID compatibility check will detect that in f_1, the background triangle is covered by another triangle (e.g., the purple one), and hence mark this location as not visible in the forward visibility mask I_t^1. The black triangle sits on the foreground object, and

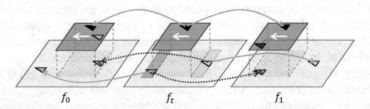

f_0 $\qquad\qquad$ f_t $\qquad\qquad$ f_1

Fig. 7.6 Triangle ID checking is used to identify regions in the target frame f_t that are not visible in either of the reference frames f_0 and f_1; we refer to the text for details

hence can be predicted from both sides, which is readily discovered by the proposed method.

As in the other methods proposed in this thesis, we use a weighted bidirectional prediction of the motion compensated reference frames $f_{0\rightarrow t}$ and $f_{1\rightarrow t}$, whenever the location **m** is visible in either *both* or *neither* of the reference frames, and switch to unidirectional prediction whenever a location is only visible in one reference frame.

$$\hat{f}_t[\mathbf{m}] = \begin{cases} (1-t)f_{0\rightarrow t}[\mathbf{m}] + tf_{1\rightarrow t}[\mathbf{m}] & I_t^0[\mathbf{m}] = I_t^1[\mathbf{m}] \\ f_{p\rightarrow t}[\mathbf{m}] & \text{otherwise} \end{cases}, \qquad (7.5)$$

where p refers to the reference frame where the location **m** is visible.

7.1.5 Qualitative Evaluation of Temporal Consistency

As mentioned earlier, interpolating temporally consistent frames becomes more important for higher framerate upsampling factors, where more frames have to be interpolated. As we mentioned in the earlier chapters, the proposed methods allow for (arbitrary) upsampling factors. However, so far, we have only evaluated the methods in the common TFI evaluation scenario of doubling the framerate, since most existing state-of-the-art TFI methods cannot easily allow for higher upsampling factors.

Assessing the temporal consistency in an objective way is an interesting research problem that is out of the scope of this thesis. In the future, we plan to develop a temporal consistency measure, which employs a lot of the reasoning we present in this thesis to provide an objective measure of temporal consistency. In order to give some insight into how well the BAM framework performs, we therefore show visual results obtained on sequences from the Sintel dataset (see Sect. A.2); Figs. 7.7, 7.8 and 7.9 show crops of three sequences, for a frame upsampling factor of 4. The development of the BAM scheme was mainly motivated by the observation that in a coding environment, hierarchically anchored motion as employed by BIHA and FOHA, leads to coding inefficiencies. In a TFI scenario, this is not an issue; in fact, the interpolation quality of the BAM and the FOA-TFI scheme is quite similar. We therefore show comparative results for the BOA-TFI scheme. It is worth noting that the BOA-TFI scheme is already tailored to create consistent results, and its performance is probably above average around moving objects. Nonetheless, the BAM framework we propose in this chapter creates more consistent results around moving objects, especially around thin objects; this is evidenced in the examples of the figures, where the yellow rectangles highlight regions where significant improvements can be observed.

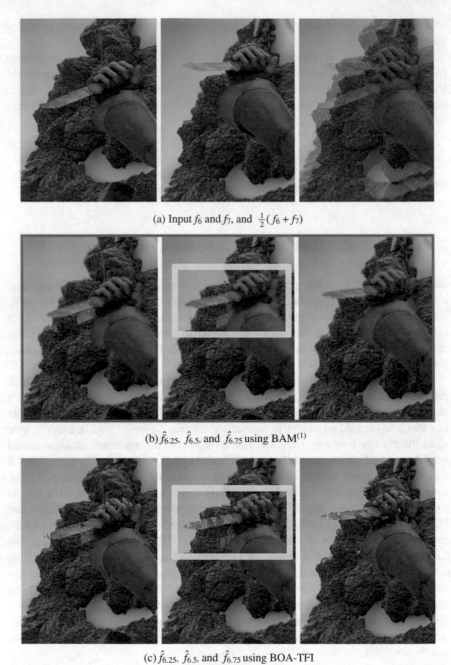

(a) Input f_6 and f_7, and $\frac{1}{2}(f_6 + f_7)$

(b) $\hat{f}_{6.25}$, $\hat{f}_{6.5}$, and $\hat{f}_{6.75}$ using BAM$^{(1)}$

(c) $\hat{f}_{6.25}$, $\hat{f}_{6.5}$, and $\hat{f}_{6.75}$ using BOA-TFI

Fig. 7.7 Comparison of the temporal consistency of BAM with BOA-TFI on the "Ambush 6" sequence, for a framerate upsampling factor of 4. **a** shows the two input frames, as well as the average of the two input frames, which gives some idea of the motion. **b** shows the three interpolated frames produced by the proposed BAM method, and **c** shows comparative results obtained using BOA-TFI

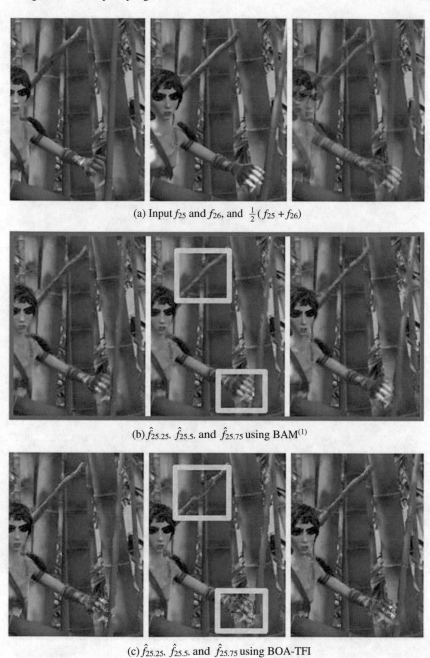

(a) Input f_{25} and f_{26}, and $\frac{1}{2}(f_{25} + f_{26})$

(b) $\hat{f}_{25.25}$, $\hat{f}_{25.5}$, and $\hat{f}_{25.75}$ using BAM$^{(1)}$

(c) $\hat{f}_{25.25}$, $\hat{f}_{25.5}$, and $\hat{f}_{25.75}$ using BOA-TFI

Fig. 7.8 Comparison of the temporal consistency of BAM with BOA-TFI on the "Bamboo 2" sequence, for a framerate upsampling factor of 4. **a** shows the two input frames, as well as the average of the two input frames, which gives some idea of the motion. **b** shows the three interpolated frames produced by the proposed BAM method, and **c** shows comparative results obtained using BOA-TFI

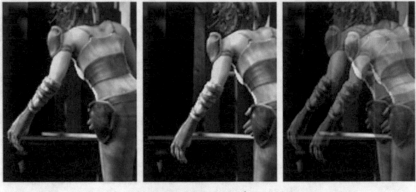

(a) Input f_{16} and f_{17}, and $\frac{1}{2}(f_{16}+f_{17})$

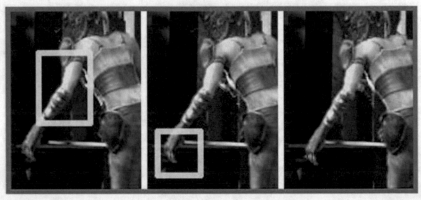

(b) $\hat{f}_{16.25}$, $\hat{f}_{16.5}$, and $\hat{f}_{16.75}$ using BAM$^{(1)}$

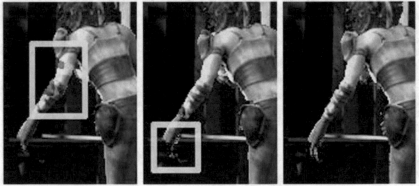

(c) $\hat{f}_{16.25}$, $\hat{f}_{16.5}$, and $\hat{f}_{16.75}$ using BOA-TFI

Fig. 7.9 Comparison of the temporal consistency of BAM with BOA-TFI on the "Market 2" sequence, for a framerate upsampling factor of 4. **a** shows the two input frames, as well as the average of the two input frames, which gives some idea of the motion. **b** shows the three interpolated frames produced by the proposed BAM method, and **c** shows comparative results obtained using BOA-TFI

7.2 Triangular Mesh Sparsification

In the earlier chapters, we employed a triangular mesh with a *fixed* triangle size of 1×1 in the CAW procedure to map motion fields from one frame to another. In regions of smooth motion, one can expect that the triangle size can be increased without significantly degrading the quality of the motion field. We employ an *indexed mesh* structure, which enables an efficient GPU implementation in the future. Besides its positive impact on computational complexity, sparsifying the motion field has other interesting benefits when it comes to compression, where it translates to compressibility. We note, however, that the mesh sparsification algorithm we propose here is not designed with R-D optimality as the objective.

Algorithm 1 shows the pseudo-code of the proposed triangular mesh sparsification algorithm. We start by partitioning $M_{0 \to 1}$ into cells of size L, where L is the largest allowed cell size. Then, each cell is split up into two triangles $T_{i,j,p,L}$, where (j, i) denote the upper left coordinates of the cell, and $p = 0$ or $p = 1$ are used to distinguish between the upper left and the lower right triangle of the cell. More precisely, the coordinates of the three vertices of a triangle $T_{i,j,p,L}$ are:

$$
\begin{aligned}
V_0^{0,T_{i,j,p,L}}(x, y) &= (j + L, i) \\
V_0^{1,T_{i,j,p,L}}(x, y) &= (j, i + L) \\
V_0^{2,T_{i,j,p,L}}(x, y) &= \begin{cases} (j, i) & \text{if } p = 0 \\ (j + L, i + L) & \text{if } p = 1 \end{cases}
\end{aligned}
\tag{7.6}
$$

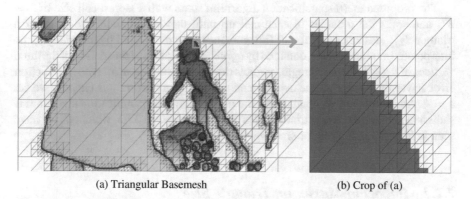

(a) Triangular Basemesh   (b) Crop of (a)

Fig. 7.10 Example base mesh created by the proposed mesh sparsification algorithm, superimposed on the (colour-coded) dense motion field. Around motion discontinuities, triangles are split up to 1×1, whereas they are larger in regions of smooth (affine) motion

---

Algorithm 1 Mesh Sparsification Algorithm

1: **procedure** CREATESPARSETRIANGULARMESH
2:    $L \leftarrow 2^N$ ▷ Start with largest cell length
3:    **for** $i = 0$ **to** $height$ **step** L **do**
4:        **for** $j = 0$ **to** $width$ **step** L **do**
5:            CREATESMOOTHTRIANGLE($T_{i,j,0,L}$)
6:            CREATESMOOTHTRIANGLE($T_{i,j,1,L}$)
7:        **end for**
8:    **end for**
9: **end procedure**
10: **function** CREATESMOOTHTRIANGLE($T_{i,j,p,L}$)
11:    **if** $\left\| \mathbf{u}[\mathbf{m}] - \mathbf{u}_{aff}[\mathbf{m}] \right\|_2 < \varepsilon \ \forall \mathbf{m} \in T_{i,j,0,L}$ **then**
12:        Create triangle $T_{i,j,p,L}$, assign unique TID
13:    **else**
14:        $L \leftarrow L/2$ ▷ Assign new cell length
15:        **if** $p == 0$ **then**   ▷ Split upper left triangle
16:            CREATESMOOTHTRIANGLE($T_{i,j,0,L}$)
17:            CREATESMOOTHTRIANGLE($T_{i,j,1,L}$)
18:            CREATESMOOTHTRIANGLE($T_{i,j+L,0,L}$)
19:            CREATESMOOTHTRIANGLE($T_{i+L,j,0,L}$)
20:        **else**   ▷ Split lower right triangle
21:            CREATESMOOTHTRIANGLE($T_{i+L,j,1,L}$)
22:            CREATESMOOTHTRIANGLE($T_{i,j+L,1,L}$)
23:            CREATESMOOTHTRIANGLE($T_{i+L,j+L,1,L}$)
24:            CREATESMOOTHTRIANGLE($T_{i+L,j+L,0,L}$)
25:        **end if**
26:    **end if**
27: **end function**

The proposed mesh sparsification algorithm starts with a largest cell size of $L = 2^N$, and splits the triangles of each cell up until they are *smooth*. We consider a triangle $T_{i,j,p,L}$ as smooth if for all (integer) locations $\mathbf{m}$ covered by the triangle, the interpolated motion $\mathbf{u}_{aff}[\mathbf{m}]$, obtained by *affine* interpolation of the motion of the three vertices of the considered triangle, predicts the original motion $\mathbf{u}[\mathbf{m}]$ with a prediction error lower than ε; for all the experimental results produced in this chapter, we set $\varepsilon = \frac{1}{2}$. An example mesh obtained by the proposed mesh sparsification algorithm is shown in (Fig. 7.10). Every triangle gets assigned a *unique* TID, which, as we have seen in Sect. 7.1.3, is used for creating visibility masks.

7.2.1 Impact of Maximum Triangle Size

The proposed mesh sparsification allows us to trade-off computational complexity and memory requirements with image reconstruction quality. For evaluation, we use the natural sequences data set used in earlier experiments (see Sect. A.3 for more details), which contains 12 sets of 21 frames each; as in the TFI experiments presented earlier in this thesis, we dropped all the odd indexed frames, which results in a total of 120 frames where a ground truth frame exists. For each even frame pair, we estimate motion using MDP-flow [2]. We choose a frame upsampling factor of 8, which interpolates 7 frames in between the two reference frames. For each frame pair, we compute the PSNR of the center frame ($t = 0.5$), where a ground truth frame exists. We note that this center frame is the hardest to predict, being equally far away from both reference frames. The closer the interpolated frame is to a reference frame, the easier the prediction gets from the respective side, and the more importance is put onto the prediction from that side, which generally results in better frame interpolation quality.

Figure 7.11 shows the impact of maximum cell size L (between 1 and 64) on processing time (solid orange line) and reconstruction quality (dashed blue line). One can observe that the maximum allowed cell size L has very little impact on the reconstruction quality; this is due to the fact that around moving objects, where the motion is expected to be less smooth, the triangles are small irrespective of the maximum allowed cell size L. In terms of processing times, one can see a significant drop in average processing time from 2.2 s for $L = 1$ (i.e., a fixed cell size of 1×1) to around 0.45 s for $L = 8$. On the tested sequences with a resolution of 832×480, increasing the maximum allowed cell size had almost no impact on processing times, indicating that the average cell size on the tested sequences was about 8. On sequences with higher resolutions, however, one can expect the knee point in the processing time plot to shift further to the right.

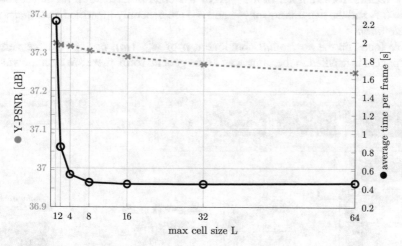

Fig. 7.11 Average per-frame processing time (solid line, in secs) and reconstructed PSNR (dashed line) as a function of maximum allowed cell size L

7.3 BAM with Higher-Order Motion

We now describe how higher order motion models can be incorporated into the BAM scheme, which, as we shall see, is particularly useful if BAM is used in a video compression framework. We remind the reader that we use BAM$^{(n)}$ to refer to the BAM scheme employing an nth-order motion model. Input to the motion modelling stage are reference frames f_0 and f_1, plus additional intermediate frames f_{t_j}; the output is an nth-order motion field, denoted as $M^{(n)}_{0 \to t}$, which describes the motion between f_0 and *any* f_t, $t \in [0, 1]$. As shown in Fig. 7.12, $M^{(n)}_{0 \to t}$ and the two reference frames f_0 and f_1 are then input to the TFI scheme.

The general form of the horizontal and vertical motion vectors following an nth-order motion model can be written as:

$$
\begin{aligned}
u^{(n)}_{0 \to t}(\mathbf{m}) &= \sum_{k=1}^{n} w_k p^{(u)}_k t^k \\
v^{(n)}_{0 \to t}(\mathbf{m}) &= \sum_{k=1}^{n} w_k p^{(v)}_k t^k
\end{aligned}
\tag{7.7}
$$

where the $p^{(u)}_k$ and $p^{(v)}_k$ denote the motion basis vectors for the horizontal and vertical components, respectively, and the w_k's are weights to scale the different basis vectors; for example, in the case of a second-order motion model, $w_1 = 1$, $w_2 = 0.5$, and the two unknowns $p^{(\cdot)}_1$ and $p^{(\cdot)}_2$ are generally identified as *velocity* and *acceleration*.

We focus on the *horizontal* component u of $\mathbf{u}^{(n)}_{0 \to t}(\mathbf{m})$, noting that the vertical component is handled in the same way. In its general form, there are n unknowns p_k,

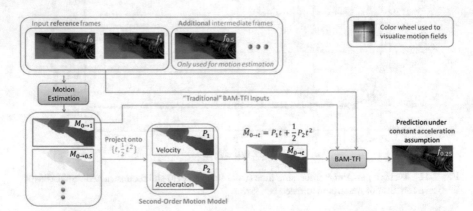

Fig. 7.12 Extension of BAM to add a second-order motion model that is able to account for accelerating motion (modifications in orange). In addition to the two reference frames, we estimate the motion between f_0 and at least one frame in between f_0 and f_1. These motion fields are then projected onto the subspaces t and $0.5t^2$, which subsequently allow us to compute $\hat{M}_{0 \to t}$ under a *constant acceleration* assumption

and hence we need at least n motion vectors $\boldsymbol{u}_{0 \to t_j}(\mathbf{m})$; this means that we require at least $n + 1$ frames as input. In matrix form, we can write the system of equations $A\mathbf{p} = \mathbf{b}$, where

$$
A = \begin{bmatrix} w_1 t_1 & w_2 t_1^2 & \cdots & w_n t_1^n \\ w_1 t_2 & w_2 t_2^2 & \cdots & w_n t_2^n \\ \vdots & \vdots & \ddots & \vdots \\ w_1 t_q & w_2 t_q^2 & \cdots & w_n t_q^n \end{bmatrix}, \mathbf{b} = \begin{bmatrix} \boldsymbol{u}_{0 \to t_1}(\mathbf{m}) \\ \boldsymbol{u}_{0 \to t_2}(\mathbf{m}) \\ \vdots \\ \boldsymbol{u}_{0 \to t_q}(\mathbf{m}) \end{bmatrix}, \tag{7.8}
$$

with $q \geq n$. From linear algebra, we know that this system of equations has a *unique* least-squares solution $\mathbf{p}^+$ of smallest norm. One way of finding $\mathbf{p}^+$ is in terms of the *pseudo-inverse* of A, denoted as A^+. That is,

$$
\mathbf{p}^+ = A^+ \mathbf{b}, \tag{7.9}
$$

which is obtained from the *singular value decomposition* (SVD) of A.

The motion parameters $\mathbf{p}^+$ can be used to create motion that follows an nth-order motion model for *any* frame f_t in between the two reference frames. For example, in the case of a second-order motion model, motion following a *constant acceleration* assumption can be obtained as follows:

$$
\boldsymbol{u}_{0 \to t}^{(2)}(\mathbf{m}) = p_1 t + 0.5 p_2 t^2. \tag{7.10}
$$

As mentioned earlier, the vertical component $\boldsymbol{v}_{0 \to t}(\mathbf{m})$ is obtained following the same development. Combining the horizontal and vertical components yields $\mathbf{u}_{0 \to t}^{(n)}(\mathbf{m})$, which is then used as input to the affine mesh mapping procedure presented in Sect. 7.1.1. Apart from the changes mentioned above, all other aspects were left "as-is" in our exploration. In the next section, we experimentally show how the prediction performance of BAM$^{(2)}$ improves over BAM$^{(1)}$.

7.3.1 Prediction Performance of Second-Order Motion Model

In this section, we employ a second motion field between each even frame and succeeding odd frame, and evaluate the improved *prediction* performance of BAM$^{(2)}$ compared to BAM$^{(1)}$.[4]

The last column of Table 7.1 shows quantitative results of the BAM$^{(2)}$ scheme, where in addition to the motion between any two *even* reference frames, we estimated motion between each even frame and the succeeding odd frame. One can see that in 8 out of the 10 tested sequences, BAM$^{(2)}$ performs better than BAM$^{(1)}$ by varying amounts. Notable improvements are achieved on sequences that contain fairly large

---

[4]Note that while the odd frame is not available in a traditional TFI framework, it is available in a coding framework, which this work builds towards.

Table 7.1 Average Y-PSNR for BAM$^{(1)}$ and BAM$^{(2)}$ on natural sequences (10 interpolated frames per sequence)

Sequence	BAM$^{(1)}$	BAM$^{(2)}$
Cactus	34.22 (−0.09)	**34.31**
Kimono1	33.53 (−0.54)	**34.07**
Kimono2	41.39 (−0.56)	**41.94**
Rushhour	34.77 (−0.73)	**35.50**
Shields1	36.20 (−0.50)	**36.70**
Shields2	37.75 (−0.13)	**37.88**
Park	**39.94 (+0.19)**	39.75
Parkrun	32.04 (−0.16)	**32.20**
Station2	43.16 (−0.10)	**43.26**
Mobcal	**38.51 (+0.13)**	38.38
Average	37.15 (−0.25)	**37.40**

regions that are accelerated, such as the "Kimono1" and "Kimono2" sequences, as well as "Rush Hour" and "Shields1". Clearly, the second-order model should be at least as good as the first-order one. However, the motion fields have to be temporally consistent in order for a higher order motion model to be meaningful, which cannot be guaranteed if the two fields are independently estimated.

As in earlier chapters, in order to mitigate the impact of suboptimal motion estimation, we further conduct a comprehensive experiment on challenging sequences from the Sintel dataset, where forward motion between adjacent frames (i.e., 1-hop motion) is available as ground truth. We construct a 2-hop motion field by concatenating two 1-hop flows; for locations that move out of the frame in the first motion field, we assume constant velocity to create the 2-hop motion vector. We drop every odd frame and use the BAM scheme to predict the odd frames. Table 7.2 shows the aver-

Table 7.2 Average Y-PSNR for BAM$^{(1)}$ and BAM$^{(2)}$ on full Sintel sequences (24 interpolated frames per sequence)

Sequence	BAM$^{(1)}$	BAM$^{(2)}$
Alley 1	31.64 (−1.62)	**33.26**
Alley 2	32.86 (−2.02)	**34.88**
Ambush 7	30.58 (−5.07)	**35.66**
Bandage 1	31.97 (−1.26)	**33.22**
Bandage 2	33.82 (−2.07)	**35.89**
Bamboo 1	29.23 (−0.06)	**29.29**
Bamboo 2	28.54 (−0.81)	**29.35**
Market 2	28.62 (−0.96)	**29.58**
Shaman 2	37.58 (−1.04)	**38.62**
Shaman 3	37.05 (−0.01)	**37.06**
Average	32.19 (−1.49)	**33.68**

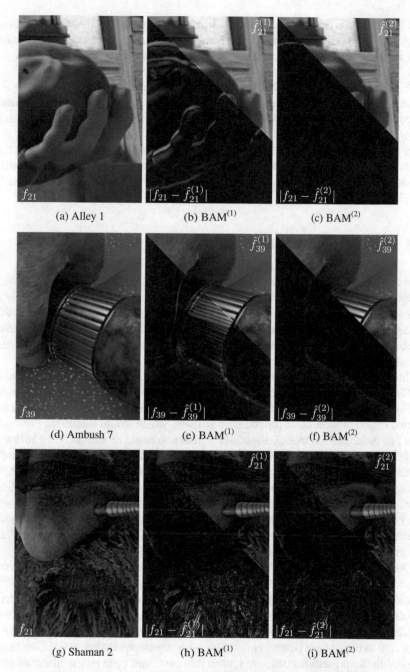

(a) Alley 1   (b) BAM$^{(1)}$ (c) BAM$^{(2)}$

(d) Ambush 7   (e) BAM$^{(1)}$ (f) BAM$^{(2)}$

(g) Shaman 2   (h) BAM$^{(1)}$ (i) BAM$^{(2)}$

Fig. 7.13 Prediction performance of the first and second order motion model on three sequences from the Sintel dataset. Each row shows the ground truth frame, followed by the predicted frame using the first-order motion and the second-order motion model, denoted as BAM$^{(1)}$ and BAM$^{(2)}$, respectively. The upper right part of the crops shows the predicted frame, and the lower left part the (absolute) difference between the prediction and the ground truth frame

age Y-PSNR on a number of sequences. In this experiment, the BAM$^{(2)}$ *consistently* outperforms BAM$^{(1)}$. Significant improvements are observed on sequences that contain large acceleration, such as the "Ambush 7" sequence, as well as "Bandage 1" and "Alley 2".

Figure 7.13 shows crops of frames from three different sequences: in "Alley 1" a hand holding an apple is moving down and slightly rotating; "Ambush 7" contains a sceptre that is heavily accelerated as it moves North-East; lastly, in "Shaman 2", the motion of the beard of the shaman is highly complex, as can be seen in Fig. A.6f. In all sequences, BAM$^{(1)}$ is able to create highly credible results (see upper right parts of the crops in the middle column of the Fig. 7.13); however, BAM$^{(2)}$ predicts the actual location of objects much better than BAM$^{(1)}$, as evidenced by the smaller prediction errors.

7.4   Video Compression Using BAM

The experiments in the previous section have shown that BAM$^{(2)}$ is able to better predict the "true" position of moving objects that are under non-constant motion. In this section, we show how the BAM scheme can be employed in a video compression system. Before we delve into the description of how this is achieved, we find it useful to compare the three motion anchoring schemes proposed in this thesis in some more details.

In Fig. 7.1, one can see how the motion anchoring progressively simplifies, leading to BAM, where all coded motion information is anchored at one frame. Note that in Fig. 7.1c, only three motion fields are coded; one full (solid black arrow), as well as two scaled motion fields (dotted blue arrows), which would enable the estimation of a 3rd-order motion model. Higher-order motion could be estimated by including additional motion fields (dotted grey arrows in Sect. 7.3c). The choice of coding motion fields $M_{0\rightarrow1}$, $M_{0\rightarrow4}$, and $M_{0\rightarrow8}$, is not arbitrary. As mentioned in Sect. 7.3, the scheme always requires the motion linking the two reference frames ($M_{0\rightarrow8}$). For fast moving objects, there might be no correspondence between f_0 and f_8, since the object could have left the frame. For this reason, it is helpful to have motion information between shorter baselines. For slow-moving objects, on the other hand, longer baselines are required to even observe the motion.

As mentioned earlier, BAM departs from the two other hierarchical motion anchoring schemes. Because of the central organization of motion information, no *inferred* motion information (dashed orange arrows in Fig. 7.1) is coded. As discussed earlier, inferred motion field residuals are expected to be nonzero only in regions of disocclusion, and only if the proposed background motion extrapolation "fails"; this is expected to be the case whenever there are new objects appearing in the disoccluded region. In the present scheme, we ignore this case, and code potential errors in the texture residual. An interesting extension to the BAM scheme would be to assess if more optimal decisions could be made by adding "accommodation" motion fields to the base mesh, which augment the motion description to describe the trajectory of such "intermediate" motion layers.

7.4.1 Integrating BAM into a Video Coder

In this section, we integrate the BAM scheme into HEVC, and provide preliminary coding results on three sequences. We estimate motion between the first frame (i.e., the base frame) and *eight* successive frames of the GOP. From these eight motion fields, a second-order motion model is estimated as described in Sect. 7.3. We then predict the seven target frames of the GOP, and code the texture residuals using HEVC; the two reference frames are coded in Intra-mode. The motion fields are coded using a modified JPEG2000 codec, which uses a breakpoint-adaptive wavelet transform to conserve motion boundaries, as used in Sect. 5.5.1. Currently, all motion fields are *separately* coded; further improvements can be expected from a *joint* motion field coding.

Figure 7.14 shows the average per-frame rate-distortion performance for three sequences, with QP-values of $\{28, 33, 38, 42\}$; the QP-offset for the target frame residuals is set to 4. We show results obtained for $\text{BAM}^{(1)}$ and $\text{BAM}^{(2)}$, as well as for HEVC. We further show the performance of a modified version of HEVC, where only the two reference frames are used as prediction references, as is the case in our framework – we refer to this modified version as "HEVC Ref Only".

On the "Station 2" sequence, which is dominated by zoom-out motion, $\text{BAM}^{(2)}$ performs on par with HEVC. The "Shields" sequence contains an inconsistent zoom, for which the (credible) constant-velocity prediction of $\text{BAM}^{(1)}$ creates large residuals, which are highly expensive to code. This is evidenced in Fig. 7.15, where we focus on one frame of the GOP. While there are hardly any visible differences between the decoded frame obtained from HEVC and the *prediction* (without residual added) of $\text{BAM}^{(1)}$, the latter results in a large prediction residual; in terms of Y-PSNR, there is a difference of more than 6dB. As can be seen in the R-D plot of Fig. 7.14b, $\text{BAM}^{(2)}$ is able to much better predict the intermediate frames, which is evidenced by the large gap between the blue and the red curve in the figure.

Finally, we tested the performance on the "Surfer Jump" sequence, which contains complex motion such as splashing waves that is not captured in the estimated motion fields. The reason that $\text{BAM}^{(2)}$ performs slightly worse that $\text{BAM}^{(1)}$ is that the *additional* motion field that has to be coded for the second-order motion model does not improve the prediction of the target frame. While HEVC outperforms the proposed scheme on this sequence, it is worth noting that the interpolated frames produced by our scheme are *highly credible*, as can be seen in Fig. 7.16. Videos of both the "Shields" and the "Surfer Jump" sequence *without residual coding* are available on our website.[5]

We end the discussion by highlighting that with the proposed framework, the framerate can easily be increased at the decoder without having to re-estimate motion, which is not possible in current video codecs. Furthermore, the centralized organization of motion should greatly facilitate ROI coding. This is because for location

---

[5]http://ivmp.unsw.edu.au/~dominicr/pcs_2016.html.

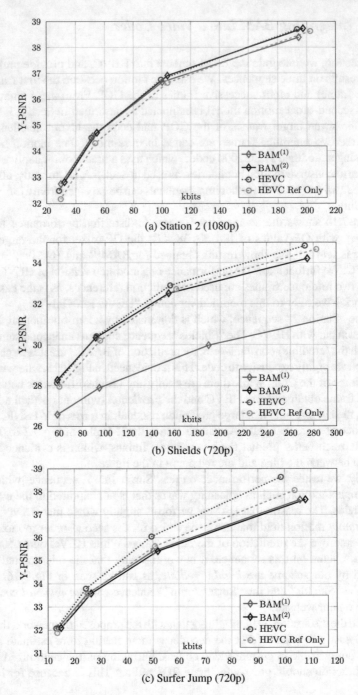

Fig. 7.14 R-D comparison of BAM with HEVC. We show R-D plots for three test sequences; the x-axis shows the average number of kbits per frame, and the y-axis the Y-PSNR

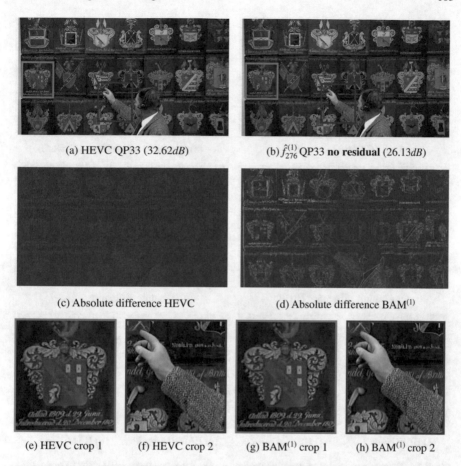

(a) HEVC QP33 (32.62*dB*)

(b) $\hat{f}_{276}^{(1)}$ QP33 **no residual** (26.13*dB*)

(c) Absolute difference HEVC

(d) Absolute difference BAM$^{(1)}$

(e) HEVC crop 1   (f) HEVC crop 2   (g) BAM$^{(1)}$ crop 1   (h) BAM$^{(1)}$ crop 2

Fig. 7.15 Visual comparison of decoded frames. **a** shows the *middle* frame f_{276} (GOP is $f_{272} - f_{280}$) from the "Shields" sequence, decoded using HEVC at QP value of 33; **b** shows the same frame produced by BAM$^{(1)}$ **without residual coding**. **c** and **d** show the absolute differences of both methods compared to the original frame; (e-h) show crops, highlighting that there is hardly any visible difference, despite the large difference in terms of PSNR

in the left base frame f_0, the trajectory through the spatio-temporal volume is completely described. Therefore, for any ROI, the information that has to be decoded is very well defined. In a traditional anchoring scheme, accessibility is much harder to achieve, since motion is organized at several frames. Additionally, the target-anchoring requires the decoding of large amounts of "unnecessary" data to discover which regions belong to the ROI.

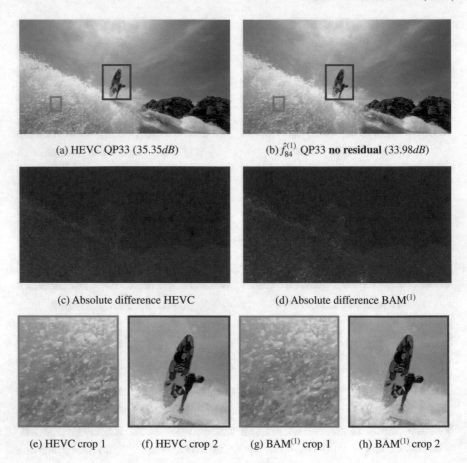

Fig. 7.16 Visual comparison of decoded frames. **a** shows the *middle* frame f_{84} (GOP is $f_{80} - f_{88}$) from the "Surfer Jump" sequence, decoded using HEVC at QP value of 33; **b** shows the same frame produced by BAM$^{(1)}$ **without residual coding**. **c** and **d** show the absolute differences of both methods compared to the original frame; **e–h** show crops of the decoded frames; while there are differences in the splashes, the difference between **e** and **g** is hardly visible

7.5 Summary

In this chapter, we presented a third reference-based motion field anchoring strategy for video compression. In this scheme, all motion information is anchored at the first frame of the GOP. This central motion organization addresses the largest shortcoming of the two hierarchical anchoring strategies presented in earlier chapters, which is the efficient composition of motion fields. We proposed a mesh sparsification algorithm that enables an adaptive cell size of the triangular motion mesh; experimental results showed that the computational complexity can be reduced by 75%, with a trivial drop in PSNR of the interpolated frames.

In regions of non-constant motion, the constant velocity assumption that is very common in TFI schemes can lead to large prediction residuals, even if the interpolated frame looks highly credible. In a compression scenario, where target frames are to be predicted, these prediction residuals can be expensive to code. With the aim of improving the *prediction* performance of the BAM framework, we incorporated higher-order motion fields, which are able to better predict the actual trajectory of moving objects through the spatio-temporal volume. Subsequently, we integrated the BAM scheme into HEVC, where initial experiments showed very promising results compared to HEVC.

References

1. D. Rüfenacht, D. Taubman, Temporally consistent high frame- rate upsampling with motion sparsification, in *IEEE International Workshop on Multimedia Signal Processing* (2016) (cited on pages vii, 143)
2. L. Xu, J. Jia, Y. Matsushita, Motion detail preserving optical flow estimation. IEEE Trans. Pattern Anal. Mach. Intell. 1744–1757 (2012) (cited on pages 43, 44, 49, 73, 95, 102, 128, 130, 131, 134, 137, 139, 159, 185–187)

Chapter 8
Conclusions and Future Directions

This thesis investigates novel motion anchoring schemes for highly scalable video compression schemes, with the aim of improving the temporal scalability and accessibility features. Existing video compression schemes employ block-based motion, which are unable to accurately represent motion information in the vicinity of moving object boundaries. Instead, in this thesis, we focus on "physical" motion, which accurately describes the trajectory of each pixel location. In a compression scenario, such dense motion fields are generally discarded because of their high coding cost. We employ a recently proposed highly scalable representation of motion discontinuities using breakpoints. Breaks can be used to adapt the wavelet bases in the vicinity of (motion) discontinuities; this significantly reduces the coding cost of motion fields while retaining sharp moving object boundaries in quantized motion fields.

Existing video compression systems describe motion at the target frames that are to be predicted. This target-based anchoring of motion implies that motion is only involved in the prediction of one frame. We show how the combination of physical motion with an explicit description of motion boundaries enables to flip the anchoring of motion from target to reference frames, which enables the re-use of motion information from coarser temporal levels at finer temporal levels, which further reduces the motion coding cost. In such a *reference-based* anchoring, motion fields have to be inverted in order to serve as a prediction references; the involved motion mapping process is one of the key steps in TFI. Observing that motion discontinuities displace with the foreground object, we propose a motion-discontinuity guided motion field inversion procedure in which double mappings are resolved in a disciplined way; furthermore, we show how motion discontinuity information can be used to identify background motion in disoccluded regions, which subsequently can be extrapolated in order to assign "physical" motion in regions of disocclusion.

In order to enable bidirectional prediction of the target frame, existing TFI schemes estimate motion information both in forward and reverse direction of the two reference frames. This doubles the amount of motion information that has to be estimated,

© Springer Nature Singapore Pte Ltd. 2018   167
D. Rüfenacht, *Novel Motion Anchoring Strategies for Wavelet-Based
Highly Scalable Video Compression*, Springer Theses,
https://doi.org/10.1007/978-981-10-8225-2_8

and – in a compression scenario – ultimately has to be coded. Instead, in this thesis we propose a powerful operation on motion fields we call motion *inference*, which can be used to obtain a *geometrically consistent* bidirectional prediction of a target frame from just one motion field linking two reference frames. Throughout the thesis, we successively refine the TFI performance; extensive evaluations and comparisons with state-of-the-art TFI methods highlight the high performance of the proposed frame interpolation scheme.

In a compression scenario, inferred motion fields are particularly appealing since they are expected to be nonzero only in regions that get disoccluded between the two reference frames, and hence are inexpensive to code. We investigate three different reference-based motion anchoring schemes in the context of (highly scalable) video compression. In the BIHA scheme, the motion anchoring of all *coded* motion fields is "flipped" with respect to the traditional anchoring of motion fields at target frames. Through careful analysis of how the spatio-temporal texture and motion subbands interact, we determine the importance of the different components, which can be used to weight the different spatio-temporal subbands. Experimental results show that the BIHA scheme outperforms the traditional anchoring of motion at target frames. The improvements are both due to cheaper motion field coding, as well as better occlusion handling and geometrical consistency, which cannot be guaranteed in a target-based anchoring if motion information is quantized.

The experimental results of the BIHA scheme are both promising and instructive; while they show some major advantages over the traditional target-based anchoring, they also allowed us to identify a number of suboptimalities. In Chap. 6, we further improve the TFI performance by addressing three key issues of the BOA-TFI scheme. First, we propose a divergence-based measure of motion discontinuity we call DFLM, which conveys a much "richer" description of motion discontinuity information than the (binary) discontinuity description induced from breakpoints that we used in the BIHA scheme. The second improvement consists of flipping the direction of the motion inference procedure, which results in a reduction of the computational complexity by roughly a factor of 3, while improving the geometrical consistency of the prediction motion fields. Finally, we propose two texture optimizations: the SWCA creates a smoother transition in regions where we change from bidirectional to unidirectional prediction, which is particularly useful in regions of illumination change. We further propose an optical blur synthesis, which creates a much smoother transition around moving objects. While both texture optimizations are targeted at improving the *visual* quality, they also have a positive impact on prediction performance; averaged over 120 interpolated frames, we observe an improvement of 0.18 dB.

In the third motion anchoring scheme proposed in this thesis, we further simplify the motion anchoring structure and anchor *all* motion information at the first frame (e.g., *base frame*) of the GOP; we call this anchoring BAM. In contrast to BIHA and FOHA, the motion anchoring of the BAM scheme is no longer *hierarchical*; as mentioned before, this does not mean that the temporal transform cannot be hierarchical. However, in this thesis, we only considered a non-hierarchical temporal transform, which allowed us to easily integrate the framework into HEVC. We believe that this

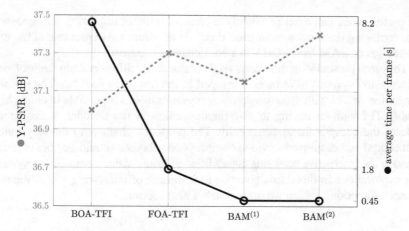

Fig. 8.1 TFI performance comparison of the three anchoring schemes proposed in this thesis. There is a clear performance increase from BOA-TFI to FOA-TFI, both in terms of quality as well as computational complexity. The mesh sparsification as employed in BAM leads to another decrease in computational complexity, with a slight negative impact on quality (BAM$^{(1)}$), which can be made up by incorporating a second-order motion model (BAM$^{(2)}$)

motion anchoring is the most promising one. The fact that all motion information is *centrally organized* has a number of advantages, including:

- Very compact motion representation. Furthermore, the fact that all motion information is coded in the same grid removes rounding errors that are inevitable in the two other, *hierarchical* anchoring strategies we proposed;
- Higher order motion models can be elegantly integrated, which improve the *prediction* performance of the TFI scheme, which is highly relevant in a compression scenario;
- ROI coding is greatly facilitated, since for any region, the trajectory through the spatio-temporal volume is known, which means that we know exactly which parts have to be decoded.

In order to improve the computational complexity of the BAM scheme, we propose a *mesh sparsification* algorithm, which uses larger triangles in regions of smooth motion. Experimental results on a large dataset show that the computational complexity can be reduced by 75%, while having a trivial impact on the PSNR.

To conclude, Fig. 8.1 shows the TFI performance (in terms of Y-PSNR) as well as processing times of the three anchoring schemes proposed in this thesis. One can see that the change from the bidirectional hierarchical anchoring (BOA-TFI) to the forward-only anchoring (FOA) leads to a significant increase in quality, while reducing the computational complexity by roughly 75%. From a TFI perspective, the FOA and BAM schemes are quite similar, with the main difference being that the BAM scheme incorporates the mesh sparsification, which reduces the computational complexity. As can be seen, there is a small drop in quality between BAM$^{(1)}$ and FOA. When incorporating a second-order motion model (BAM$^{(2)}$), however, the performance can even be slightly increased, without impacting the processing time (ignoring the motion estimation time). In addition, we expect one of the main advantages of BAM over FOHA in a compression scenario.

The ideas presented in this thesis are fundamentally different from the way video compression systems have been developed in the last three decades. The seamless integration of TFI with video compression systems makes it possible to obtain high-quality TFI without having to re-estimate motion at the decoder, which greatly reduces the computational complexity. The proposed changes to the way motion is anchored and employed in video compression schemes should be able to greatly increase the interactive browsing capabilities of future video compression systems. The explorations in this thesis give rise to a number of interesting and challenging research directions, which we present in the next section.

8.1 Future Research Directions

We conclude this thesis with an outline of a number of interesting research questions. **Tailored Motion Estimation Schemes** The quality of the TFI schemes proposed in this thesis depends on the quality of the motion fields that are employed. Ideally, motion fields should exhibit the following characteristics:

1. Piecewise-smoothness with discontinuous jumps at moving object boundaries;
2. Temporal consistency across frames; that is, the motion discontinuities of a motion field $M_{i \to j}$ should map to corresponding motion boundary information at frame f_j;
3. Multiple motion fields anchored at the same frame should share one set of motion discontinuity information;

State-of-the-art optical flow estimators are able to produce suitable results in terms of the first characteristic. The other two points, however, are specific requirements that emerge from the proposed motion inference schemes, and are therefore not considered in existing optical flow estimators. Especially the last point is hard to be satisfied if motion fields are independently estimated. For this reason, the coding results on natural sequences in Sects. 5.5.4 and 7.4.1 are limited to sequences that contain relatively few discontinuities. In order to be able to perform more comprehensive evaluations on natural test sequences, a *joint* motion estimation scheme with a *discontinuity alignment constraint* needs to be developed.

Integration of FOA-TFI and BAM-TFI into Scalable Compression Systems In Chap. 5, we presented a detailed integration of the BOA-TFI scheme into a highly scalable video compression system. For FOA-TFI, we focused on the TFI performance of a simplified motion anchoring scheme, which significantly improved over the interpolation performance of the BOA-TFI scheme. We have outlined how FOA-TFI could be integrated into a highly scalable video compression system, which we refer to as FOHA, but left the actual integration and experimental validation for future work. While there is little doubt that the R-D performance of the FOHA scheme would improve over the one of the BIHA framework, it is unclear by how much.

In Sect. 7.4.1, we integrated the BAM scheme into HEVC, and tested its performance in a single-layer compression scenario. The prime reason for this was that it allowed us to focus on the frame interpolation quality, while at the same time benefiting from the highly optimized compression pipeline provided by HEVC. The next logical step is to implement BAM in a highly scalable video compression system. This implementation should be quite straight-forward; however, in order to be able to perform a meaningful comparison, the motion estimation scheme first has to be improved, as detailed above.

Improving the Motion Extrapolation Procedure in Disoccluded Regions The proposed motion extrapolation techniques are applied on a per-triangle basis. As mentioned in Sect. 5.5.1, this can lead to artificial high-frequency content in the mapped motion fields at triangle boundaries, which are expensive to code; this is evidenced in the "Flower" sequence (see Table 5.2), where the *inferred* motion fields become very expensive to code. In the BAM scheme, this is not an issue since no mapped motion information serves as prediction reference. However, the artificial discontinuities in the mapped motion fields can lead to high-frequency content in the motion-compensated prediction of the texture information. Therefore, it would be beneficial to have a more global approach at interpolating motion in disoccluded regions.

In the BAM scheme, this could be achieved by adjusting the motion of V^{SPLIT_BG} vertices (see Sect. 7.1.2 and Fig. 7.5) so as to minimize the overall folding energy across triangle edges, before they are mapped back to the base frame to form local background motion layers.

Temporal Consistency Measure The central organization of the BAM scheme facilitates the enforcement of the temporal consistency of the interpolated frames. In particular, regions that get disoccluded between the two reference frames get assigned temporally consistent motion. However, there exists no quantitative measure for assessing temporal consistency, which is why we had to resort to qualitative comparisons for the evaluation of the temporal consistency in Sect. 7.1.5. The development of such a temporal consistency measure would be particularly useful for high framerate-upsampling-factors, as it would be more informative than (individually computed) PSNR values for the upsampled frames. Besides of being useful for comparing the temporal consistency of different TFI schemes, this measure could be directly integrated into the BAM scheme to further improve the overall consistency of the interpolated frames.

8.2 Final Remarks

The motion anchoring strategies explored in this thesis represent a fundamental change to the way motion is employed in a video compression system – from a "prediction-centric" point of view to a "physical" representation of the underlying motion of the scene. The proposed "reference-based" motion anchorings can support computationally efficient, high quality temporal motion inference, which requires as few as half of the coded motion fields compared to conventional codecs. This raises the prospect of achieving lower motion bit-rates than the most advanced conventional techniques, while providing more temporally consistent and meaningful motion. The availability of temporally consistent motion can facilitate the efficient deployment of highly scalable video compression systems based on temporal lifting, where the feedback loop used in traditional codecs is replaced by a feedforward transform.

The novel motion anchoring paradigm proposed in this thesis is well-adapted to seamlessly supporting "features" beyond compressibility, including high scalability, accessibility, and "intrinsic" frame upsampling. These features are becoming ever more relevant as the way video is consumed continues shifting from the traditional broadcast scenario with predefined network and decoder constraints to *interactive browsing* of video content over heterogeneous networks.

Appendix A
Video Test Sets

In the following, we present the different video test sets that we use throughout the thesis for evaluation purposes. The three test sets are:

1. Own synthetic sequences, where ground truth motion between *any* pair of frames in a GOP is known;
2. The Sintel dataset is a synthetic dataset that contains a large variety of highly challenging sequences with complex motion. For this dataset, 1-hop forward motion is known;
3. Common natural test sequences. For these sequences, motion fields have to be estimated.

Before we show sample frames and motion fields, we give an example of the motion field visualization we use throughout the thesis in Fig. A.1.

Fig. A.1 Visualization of motion fields using a colour-code. The direction and (normalized) magnitude of the motion vector at each location can be read from the colour in the colour-wheel; the more saturated, the larger the magnitude of the motion

© Springer Nature Singapore Pte Ltd. 2018
D. Rüfenacht, *Novel Motion Anchoring Strategies for Wavelet-Based Highly Scalable Video Compression*, Springer Theses,
https://doi.org/10.1007/978-981-10-8225-2

A.1 Synthetic Sequences

See Figs. A.2 and A.3.

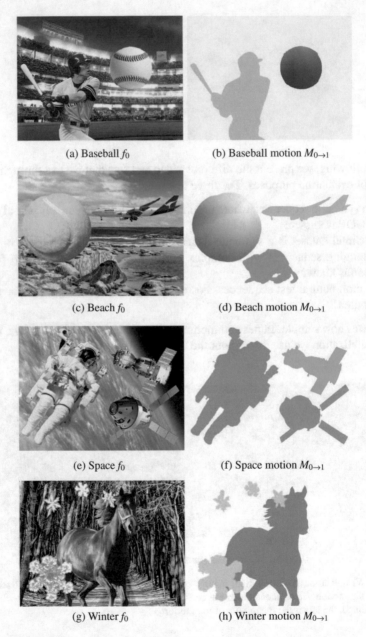

(a) Baseball f_0

(b) Baseball motion $M_{0\to1}$

(c) Beach f_0

(d) Beach motion $M_{0\to1}$

(e) Space f_0

(f) Space motion $M_{0\to1}$

(g) Winter f_0

(h) Winter motion $M_{0\to1}$

Fig. A.2 Own synthetic sequences part 1/2. Full sequences and motion fields are available on http://ivmp.unsw.edu.au/~dominicr/biha_scheme.html

(a) Autumn f_0

(b) Autumn motion $M_{0\to1}$

(c) Balls f_0

(d) Balls motion $M_{0\to1}$

(e) Butterfly f_0

(f) Butterfly motion $M_{0\to1}$

(g) Flowers f_0

(h) Flowers motion $M_{0\to1}$

(i) Robots f_0

(j) Robots motion $M_{0\to1}$

Fig. A.3 Own synthetic sequences part 2/2. Full sequences and motion fields are available on http://ivmp.unsw.edu.au/~dominicr/biha_scheme.html

A.2 Sintel

See Figs. A.4, A.5 and A.6.

(a) Alley 1 f_{14} (b) Alley 1 motion $M_{14\rightarrow15}$

(c) Alley 2 f_{14} (d) Alley 2 motion $M_{14\rightarrow15}$

(e) Ambush 7 f_{14} (f) Ambush 7 motion $M_{14\rightarrow15}$

(g) Bamboo 1 f_{14} (h) Bamboo 1 motion $M_{14\rightarrow15}$

(i) Bamboo 2 f_{14} (j) Bamboo 2 motion $M_{14\rightarrow15}$

Fig. A.4 Sintel Sequences part 1/3. Full sequences are available on http://sintel.is.tue.mpg.de/downloads

(a) Bandage 1 f_{14}

(b) Bandage 1 motion $M_{14\to15}$

(c) Bandage 2 f_{14}

(d) Bandage 2 motion $M_{14\to15}$

(e) Market 2 f_{14}

(f) Market 2 motion $M_{14\to15}$

(g) Market 5 f_{14}

(h) Market 5 motion $M_{14\to15}$

(i) Market 6 f_{14}

(j) Market 6 motion $M_{14\to15}$

Fig. A.5 Sintel Sequences part 2/3. Full sequences are available on http://sintel.is.tue.mpg.de/downloads

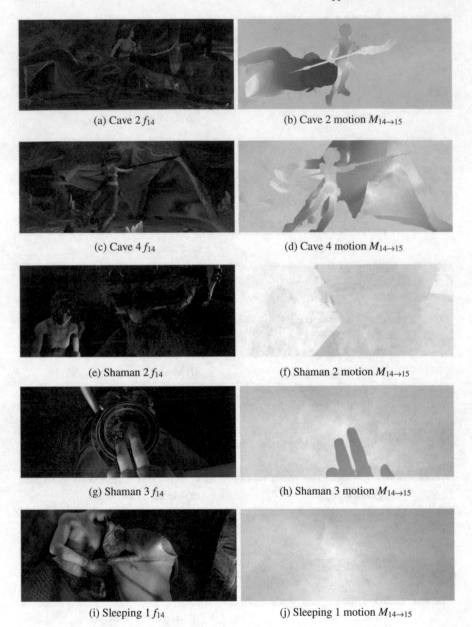

(a) Cave 2 f_{14} (b) Cave 2 motion $M_{14\rightarrow15}$

(c) Cave 4 f_{14} (d) Cave 4 motion $M_{14\rightarrow15}$

(e) Shaman 2 f_{14} (f) Shaman 2 motion $M_{14\rightarrow15}$

(g) Shaman 3 f_{14} (h) Shaman 3 motion $M_{14\rightarrow15}$

(i) Sleeping 1 f_{14} (j) Sleeping 1 motion $M_{14\rightarrow15}$

Fig. A.6 Sintel Sequences part 3/3. Full sequences are available on http://sintel.is.tue.mpg.de/downloads

A.3 Natural Sequences

See Figs. A.7, A.8 and A.9.

(a) Cactus f_6

(b) Cactus motion $\hat{M}_{6\rightarrow8}$ [†]

(c) Kimono1 f_0

(d) Kimono1 motion $\hat{M}_{0\rightarrow2}$ [†]

(e) Kimono2 f_{174}

(f) Kimono2 $\hat{M}_{174\rightarrow176}$ [†]

(g) Mobcal f_{392}

(h) Mobcal $\hat{M}_{392\rightarrow394}$ [†]

Fig. A.7 Common natural sequences part 1/3. Full sequences are available on https://media.xiph.org/video/derf. [†]Motion estimated using MDP-Flow [1]

(a) Park f_{138}

(b) Park motion $\hat{M}_{138\to140}$ †

(c) Parkrun f_{146}

(d) Parkrun motion $\hat{M}_{144\to146}$ †

(e) Shields1 f_{100}

(f) Shields1 $\hat{M}_{100\to102}$ †

(g) Shields2 f_{384}

(h) Shields2 $\hat{M}_{384\to386}$ †

Fig. A.8 Common natural sequences part 2/3. Full sequences are available on https://media.xiph. org/video/derf. †Motion estimated using MDP-Flow [1]

(a) Station2 f_{20} (b) Station2 motion $\hat{M}_{20\to22}$ †

(c) Terrace f_{138} (d) Terrace motion $\hat{M}_{138\to140}$ †

(e) Rushhour f_{48} (f) Rushhour $\hat{M}_{48\to50}$ †

(g) Stockholm f_{260} (h) Stockholm $\hat{M}_{260\to262}$ †

Fig. A.9 Common natural sequences part 3/3. Full sequences are available on https://media.xiph. org/video/derf. †Motion estimated using MDP-Flow [1]

Reference

1. L. Xu, J. Jia, Y. Matsushita, Motion detail preserving optical flow estimation. IEEE Trans. Patt. Anal. Mach. Intell. **34**, 1744–1757 (2012) (cited on pages 43, 44, 49, 73, 95, 102, 128, 130, 131, 134, 137, 139, 159, 185–187)

Printed in the United States
By Bookmasters